Smart Inventory Solutions

Improving the Management of Engineering Materials and Spare Parts

Second Edition

Phillip Slater

Industrial Press Inc.
New York, New York

Library of Congress Cataloging-in-Publication Data

Slater, Phillip.

Smart Inventory Solutions: Improving the Management of Engineering Materials and Spare Parts/ Phillip Slater.

p.cm.

ISBN 978-0-8311-3401-3

1. Inventory Management.   2. Maintenance.   I. Title

HD40.S57 2010

658.7'87—dc22

2010049777

Industrial Press, Inc.

989 Avenue of the Americas

New York, NY 10018

Printed in the United States of America

10   9   8   7   6   5   4   3   2   1

Dedication

Ask any author about the process of writing a book and I am certain that they will tell you that it is hard work.   Researching, writing, and trying to eliminate ambiguity are all more difficult than it might seem. However, hard as it is for the author, it is undoubtedly harder for the family of the author.

By its nature, writing is a solitary pursuit. Add in the constant travel when working with clients and the pressure of deadlines and an author can be become very hard to live with!

For her patience, understanding and turning of the other cheek, I dedicate this book to my darling wife, Mercedes.

Praise for Smart Inventory Solutions, Second Edition

This is a well-written, practical book. Phill Slater speaks from experience, and shows that critical inventory scrutiny will bring real bottom line savings. His message is very relevant to our studies, and I will be recommending this book to our students.

Ray Beebe,
Senior Lecturer,
Postgraduate Programs in Maintenance and Reliability Engineering
Monash University

Phillip Slater is a recognized leader in his field and has developed a proven system for reviewing and rationalizing spares. Smart Inventory Solutions is insightful, practical, and sometimes controversial and I thoroughly recommend this book to anyone that has an interest in reducing their materials and spare parts investment or who just wants to improve their materials and spares parts management.

Steve Turner
PMO Specialist
OMCS International

Smart Inventory Solutions is packed with practical ideas on how to improve management of inventory "The Forgotten Investment!". The book not only provides enlightening concepts but also a "How to" approach for accomplishing tasks. Philip Slater makes eloquent arguments for the important role that inventory needs to play in today's competitive environment.

Zahra Jabiri, PhD, MEng, BEng
Adjunct Faculty
Asia Pacific International College

I have worked in the airline industry for many years and lived through the traps that are described herein. Aircraft parts are extraordinarily expensive, so mismanagement of inventory can easily cripple an operation. Phillip presents the philosophy of real inventory optimization in a clear, logical and pragmatic manner, so that everyone in the materials management chain can understand and be part of the process. Following the steps he has detailed will certainly give a continuing competitive advantage to any aircraft operator, which could be the tipping point for flourishing, when others are just surviving in this extremely volatile and cutthroat industry. I wish I had read this book years ago!

Thomas H. Carroll III
Director of Maintenance Technical Services
NetJets® Services

Also by Phillip Slater

A New Strategy for Continuous Improvement: 10 Steps to Lower Costs and Operational Excellence (Industrial Press, 2006) ISBN 978-0-8311-3320-7

Contents

Index of Tables and Figures

Chapter 10: Execution: Taking Action to Achieve Results

Chapter 11: Case Studies

Phillip Slater

Phillip Slater is a leading authority on materials management and specifically, engineering spares management. He regularly speaks at Engineering and Asset Management conferences, writes articles for magazines worldwide, and provides services to a diverse range of clients globally. He has also authored several books.

With nearly 25 years of operational improvement experience Phillip Slater provides a broad combination of skills and experience.

Phillip has been leading organizational turnarounds in maintenance and operations since he was 24 years of age and, as a result, was rapidly promoted through maintenance into operations and then executive management.

A desire to 'do his own thing' led Phillip to leave it all behind in 2001 to start his own company, Initiate Action. Over the years Phillip has helped many companies improve profits, increase reliability, and reduce working capital in many different industries and at least a dozen different countries.

Now Phillip utilizes the expertise and experience built up over nearly 25 years to assist select clients in operations improvement and inventory optimization.

Through his company, Initiate Action, Phillip also licenses his intellectual property and processes to other companies for software development and service delivery.

You can find further details on Phillip's services at www.PhillipSlater.com.

Foreword to the Second Edition

Albert Einstein is reported to have said 'The important thing is to not stop questioning.' The most obvious question to ask when considering a second edition of a book is: Didn't you cover this topic in the first book? The answer to this question is both yes and no.

Yes, because the first book covered the Inventory Cash Release® Process and has subsequently helped thousands of people better understand that process.

No, because, through application, processes evolve and the needs of the readers evolve with them. It seems that publication of the first edition of Smart Inventory Solutions raised as many questions as it answered!

Question like:
- How does this relate to how my store room functions?
- Where does classic inventory management theory come in?
- What steps do I take next?
- How do I recognize a good consignment deal?

And so on.

It is now more than four years since I wrote the first edition of Smart Inventory Solutions and during that time I have continually answered the follow-up questions that have been asked by readers, workshop attendees, and clients. This has resulted in a whole body of work that was not widely published. As this body of work related directly to the issues raised and processes discussed in the first edition, it seemed an obvious move to publish an updated second edition.

This second edition, therefore, is not just a minor review but a complete rewrite with many new sections, tables, and diagrams — in fact, the content has nearly doubled.

Like the peeling of an onion, I fully expect that this edition will produce still further questions; I welcome any feedback from readers on issues and queries that this book provokes for them.

Keep on improving,

Phillip Slater

Chapter 1

Introduction to Smart Inventory Solutions

Achieving True Inventory Optimization

The optimization of materials and spare parts holdings has, for many years, been a goal of inventory managers, operations managers, and financial managers. This is particularly so in industries with significant investments in these types of inventory and is often driven by both a need for operational support and a need to free up working capital. The benefits that come from achieving materials and spares inventory optimization are such that the business case for action is irrefutable.

The problem is that, until now, the techniques that have been applied to achieve the optimization of this inventory do not result in a truly optimal outcome. At best they provide short-term changes; at worst they lull people into a sense of operational excellence when nothing could be further from the truth. When subject to critical analysis, the reasons for this become quite obvious. Let me explain.

Traditionally, one of two approaches has been applied: either undergo data analyses using a statistically-based software algorithm, or select a single criterion (such as obsolescence) and look for inventory that meets that criterion. Both of these approaches overlook one simple fact.

High levels of materials and spares inventory are a symptom of the broader issues with the way the inventory is controlled, supplied, accessed, purchased, and managed. Inventory levels are determined by a combination of supply chain management, internal policies and processes, and people's behavior and training. These issues involve a wide range of personnel who

1

come from engineering, maintenance, stores, inventory management, procurement, and even finance. Achieving true inventory optimization and lasting results requires an understanding of all the behaviors, context, and process factors that influence the inventory. Addressing high levels of materials and spares inventory through traditional inventory review and 'optimization' techniques does not address these issues as it rarely involves the range of personnel that influence the outcomes and does not improve the company systems and knowledge to address these issues in the future.

Neither software algorithms nor single criterion solutions address the full range of possibilities and so they cannot possibly provide true optimization. The short-lived gains they do achieve will, most likely, be reversed when the focus is taken off the immediate project and the operational processes and behaviors once again influence the inventory levels.

Achieving true inventory optimization requires a new, innovative approach that combines knowledge of parts usage, procurement, and supply chain issues with a review of behaviors and the management processes that drive them. This approach is known as the Inventory Process Optimization™ Method.

This second edition of Smart Inventory Solutions expands on the key material of the first edition to encompass and explain the Inventory Process Optimization™ Method. In doing so it also explores the supply chain, policy, process, people, and behavior issues that must be addressed for the achievement of true inventory optimization.

Inventory: The Forgotten Investment

Materials and inventory management is one of the most important disciplines in almost every company. Inventory can provide the capability to fulfill a customer need, repair a broken machine, assemble products for sale, or just keep production going. Yet, inventory management is widely perceived to be one of the most boring management topics there is. Mention inventory management to most people and almost immediately their eyes begin to glaze over. In fact, someone once suggested to me that the title of this book should be: Inventory: More Exciting Than It Sounds!

However, consider this: for manufacturing organizations, inventory can account for up to 50% (or more) of the current assets of the business (see

Appendix A for a glossary of terms used in this book). This means that for most manufacturers, up to 50% of their assets that could convert to cash in the next 12 months are tied up in inventory. For retail and wholesale businesses, the figure is even higher. While materials and spares holdings typically do not reach these values in terms of percentage of assets, a large number of organizations have millions and even tens of millions of dollars tied up in this type of inventory. Some larger organizations with which I have worked even have hundreds of millions of dollars tied up in spares.

The problem is that, unlike cash, the money tied up in this type of inventory is not available for any other use. It cannot be used to fund the business or for further investment in other productive assets. With engineering spares, this problem is even more exaggerated as the inventory is not purchased with resale in mind. The funds spent on this type of inventory are essentially gone; there is little probability of a genuine return on investment. Unlike receivables, this inventory does not represent a defined future stream of income. (The exception to this issue of resale revenue is the combination of wholesalers and final suppliers of parts who do sell their inventory. However, the ideas and processes in this book do apply equally to them.)

Despite these obvious financial issues, and the quantum of funds that are invested in this inventory, engineering materials and spares inventory management isn't a serious business topic for many people. This inventory is something that trades people and engineers stress over, accountants count, or stores people store. Many people in business concern themselves only with strategy or sales or process management or IT solutions because these are seen as high profile and 'sexy.' At the other extreme, some people consider inventory only as a means to an end. The attitude is to stock more 'just in case' but the cash impact of this is not always fully appreciated. Inventory management, it seems, is considered by many to be an activity at too low a level to create genuine financial advantage because changes in this area don't directly affect operating budgets or profit. By definition, the working capital tied up in this inventory is a cash expense that does not appear on the 'Profit and Loss' statement. Therefore, it gets little attention.

This is why engineering materials and spare parts really are the forgotten investment.

Inventory Process Optimization™

The Inventory Process Optimization™ Method aims to cast inventory in a different light. Taking a proactive approach to optimizing processes can provide significant financial advantage and enable companies to free up millions of dollars in cash. This is money that has been invested in inventory, but which either wasn't needed in the first place or is no longer needed due to a change in the market or operating environment. These changes could include a change in the level of demand, a change in the ability to supply, or both. In either case, there is an opportunity to free up cash and make alternative investments.

This book addresses the range of issues faced when managing inventory and details an approach to inventory review and reduction that has been proven to work and deliver results. Using the processes presented in this book, one company recently achieved an $18M inventory reduction in a little over 14 months. This represented a 36% reduction in the total value of their inventory. Importantly, they generated a $24M improvement in cash flow which they then invested in business improvements projects. These benefits were achieved with no loss in operational integrity, no capital investment, and no significant use of external resources. There are few investments of time or money that can match that kind of return!

By following the process and actions set out in this book, you too can implement some smart inventory solutions and achieve true inventory optimization.

Management, Reduction, or Optimization?

In all fields of management, the language and terms that are used evolve and change over time. Sometimes these changes reflect a broadening in expectations, such as the evolution of 'maintenance' into 'asset management.' Sometimes it is an innocent confusion that results from widespread use. And sometimes, it is a deliberate obfuscation by people in order to gain advantage by making the simple look complex or vice versa.

In my experience the terms inventory management, inventory reduction, and inventory optimization are frequently interchanged and misrepresented. For the sake of clarity, the following are the ways in which these terms are used in this book. Understanding these terms in context will help take us on the first step towards understanding why process optimization is the real issue in improving inventory outcomes.

Materials Management

Think of Materials Management as the top level process for control of all materials. This includes the recognition of need, estimating requirements, purchasing, and logistics of supply. Some materials may be purchased for immediate use and so will not end up as inventory

Inventory

Inventory can be defined as: All materials and spare parts that are held for future use without knowing exactly where and/or when the item will be used. This definition is discussed further in Chapter 2.

Inventory Management

Inventory Management is the activity that ensures the availability of inventory items in order to be able to service internal and/or external customers. In an operational environment, the customer will be the maintenance and production departments; in a finished goods environment, the customer is the external customer. Inventory management involves the coordination of purchasing, manufacturing, and demand to ensure the required availability.

As inventory management aims to ensure availability, the focus, almost invariably, is on the minimization (or even elimination) of stockouts. That is, minimizing any occurrence where there is not sufficient stock to meet demand. The logic that drives this behavior is as follows: Running out of inventory has consequences and there always seems to be a need for blame. Being blamed for something is an unpleasant experience for most people and in extreme cases can be seen as 'career limiting.' Therefore, any stockout triggers an action, not only to restock but also typically to overstock, in order to avoid the negative consequences of the stockout. If the inventory is already overstocked, for whatever reason, there may not be a stockout to trigger a need to take action. Therefore, there is no signal that a problem exists. A company may be overstocked and never perceive that it has a problem. Hence, a specific program of activity is required to identify these items so that their stocking can be adjusted to more appropriate levels. The result of this approach is overstocked inventory.

From this description we can already see that the traditional approach to inventory management involves processes that drive behaviors that systematically overstock inventory. This is why so many companies are overstocked.

Right now, I know that the purists will be saying, 'Wait a minute, we can put in signals to indicate overstocking such slow-moving stock indicators.' However, the practical reality is that these flags are rarely acted upon until the overall value becomes too large to ignore!

Inventory Optimization
Inventory optimization is an analytical technique that uses historical data and theoretical formulae to calculate the required level of inventory for a desired level of availability. Inventory Optimization can be a very attractive approach because it is 'fact based.' However, this strength is also its weakness because quality data is so hard to achieve (more on this in Chapter 9).

Also, by its very definition, this approach must assume that all of your existing conditions and processes are fixed; otherwise, the calculations cannot be completed. This means that Inventory Optimization does not and, in fact, cannot address the process and behavior issues that actually drive your inventory outcomes. For these reasons, Inventory Optimization can never be a solution to your inventory problems. It can only ever be a tool that is used as part of a wider program of review and then it must be used with caution.

Inventory Reduction
Inventory reduction really is just a goal. It is not a technique or a process. For example, retailers many have an 'Inventory Reduction Sale' where they sell unwanted stock at low prices. With engineering materials and spares inventory, the goal might be to reduce the level of investment in inventory (that is, the working capital or cash that is tied up) without negatively impacting the operational results.

Inventory Process Optimization™

Inventory Process Optimization™ addresses all of the shortcomings from the above. By combining inventory management fundamentals with optimization techniques, and utilizing systems thinking (Chapter 4), double loop learning (Chapter 8), and hypothesis driven analysis (Chapter 8), this approach systematically addresses process and behavior issues while identifying the specific inventory items to work on to achieve the goals of inventory reduction.

Inventory Process Optimization™ challenges the constraints to improvements in inventory to ensure that the result is an improvement to 'what could be,' not just a recalculation of 'what is.' So, while, Inventory Management must eventually lead to an over-investment of cash in inventory as people seek to eliminate stockouts, Inventory Process Optimization™ results in a sustainable and lasting minimization of the investment of cash while maintaining the availability promise. Although this difference may seem subtle, the impact is significant. Chapter 9 of this book takes you, in detail, through the Inventory Process Optimization™ Method.

Generating Lasting Results

Modifying your systems to improve your materials and spares inventory management performance is of little value if the results don't last. In fact, the use of resources in a program that doesn't deliver lasting results could be viewed as a waste of company resources. Similarly, an inventory reduction in one year is of only minor value if the cash is then put back into inventory in the next year — this is sometimes referred to as a 'yo-yo' effect. That is, great initial results and then, without other changes in the management of inventory, the initial results are reversed as the inventory is restocked. The yo-yo effect is typically the outcome of attempts at inventory reduction that don't focus on achieving lasting results through identifying and changing the real drivers of high inventory levels.

Ensuring that the gains achieved are lasting and sustainable requires the training of your team, consistent application of the process, a review of policies, procedures, metrics and measures, and changes in the reporting used for inventory management. It is only by following this path that a company-wide, lasting result can be achieved.

To assist you in achieving lasting results, Chapter 10 works through the execution of a program of inventory review and Chapter 11 works through some case studies showing short- and long-term results as well as some of the hurdles to implementation. Achieving lasting results requires more than just an understanding of Inventory Process Optimization™ techniques and processes. It also requires an understanding of the management systems that drive your inventory outcome and the implementation and tracking of the actions that you will develop from reading this book.

Henry Ford once said, 'If you think you can do something and if you think you can't, in either case you are probably right.' If you genuinely believe that no change is possible, then you are unlikely to drive a successful result. Remember, though, that to believe that no change is possible is to believe that your system is perfect and unchangeable. How likely is it that this is true?

If you recognize that past assumptions may no longer be valid, that the world changes, and that your own efforts in supply chain or operational improvement have changed the dynamics of supply and demand, then the Inventory Process Optimization™ Method is for you.

This book has been designed and written to provide you with the insight, understanding, process, and tools to improve your materials and spares inventory management performance, free up significant cash, and achieve lasting results.

Chapter 2

The Mechanics of Inventory Management

Simple not Simplistic

The purpose of this section is not to provide an in-depth understanding of all the mechanics of materials and inventory management (there are hundreds of other books that can do that) but rather to revisit or recap the mechanics in a way that puts the rest of this book in context.

Certainly this section addresses some basic elements of materials and inventory management but, as in many situations, the real value comes from looking beyond the basic element and understanding the implications of that formula or process that the element represents. Anyone can look up a formula on the Internet; however, it is an understanding of the nuances that provides value. For example, one of the major reasons that organizations have so much difficulty with materials and inventory management is that they confuse simplistic with simple. Although the concepts and formula are simple — in that they are readily understood — they should not be thought of or applied in a manner that is simplistic, that is, free of any complicating factors.

Too often people take the basic concepts and formula and do not think through the complicating factors when implementing the concept or formula. This does not mean that we need to make things overly complex; it just means that we need to understand the implications of our choices. This section aims to provide some insight to assist in that understanding.

Materials or Inventory?

One of the most common confusions of terms is that between materials and inventory. Many engineers, in particular, concern themselves with materials management but show little interest, and sometimes don't want to be involved in, inventory management. However, when subject to some scrutiny this does not make any sense.

Materials are the physical parts and components and materials management is concerned with the logistics associated with the procurement and supply of those materials in a timely manner. Inventory is the gathering of those materials in order to provide a ready supply when they are needed. Inventory can also be defined as those parts purchased for future use without knowing exactly where and/or when the part may be used.

Inventory management, therefore, is a subset of materials management as it is primarily concerned with managing the procurement and supply of those materials held as inventory. Anyone who is involved in materials management must therefore have an interest in inventory management if their materials are being held for future use. Inventory management is much more than just managing the storeroom.

Relevant Terms and Expressions

Before working through the rest of this book and to ensure a common understanding of some relevant terms and expressions, review Table 2-1, which shows the most immediately relevant terms and expressions, and Appendix A, which contains a comprehensive glossary of materials and inventory management terms and expressions.

Term	Definition
Cash flow	The net gain or loss of cash in a business through its business cycle.
Inventory	Materials and spare parts that are held for future use without knowing exactly where and/or when the item will be used.
Lead Time	Measured from when the ROP is reached to the actual physical restock.
Materials	All items that are purchased for use in, or for supporting, manufacturing and engineering activities.
Max	In some systems, this is used to determine the reorder quantity when the minimum is reached.
Min	In some systems, this is both the safety stock level and the reorder point.
Reorder Point	The trigger point for reordering stock (ROP).
Reorder Quantity	The quantity to be reordered when the ROP is reached (ROQ).
Safety Stock	An allowance for both demand and supply variations during the lead time to restock.
Stock Keeping Unit	Refers to any individual inventory item; also known as an SKU.
Stock turn	The number of times in a year that, in theory, the inventory is completely repurchased. Higher is better.
Storeroom	The area for storing the inventory; sometimes referred to as the warehouse or the store.
Stores	Sometimes used as a synonym for inventory.
Working Capital	The cash invested in inventory.

Table 2-1: Relevant Terms and Expressions

What is the Purpose of Having Inventory?

Before moving on to develop an understanding of how to manage and optimize inventory, it must first be recognized that inventory exists for a purpose. In fact there are three main reasons that companies invest in inventory:

1. To enable supply in a timely manner.
    Companies invest in inventory to ensure that the item is available when needed. The investment is based on an assessment that without the inventory there will be a loss of profit due to missed sales. This is true even for spare parts where equipment downtime may ultimately result in a lost product sale — obviously not all customers are willing to wait for delivery.

2. To provide purchasing/manufacturing efficiencies.
    Order quantities and batch sizes sometimes force companies to buy more than they need or want in the short term. In this case, they invest in the inventory in order to avoid the extra cost associated with small quantities or batches. (See also Chapter 6: Issues, Myths, and a Few Home Truths.)

3. As a temporary measure to accumulate stock prior to major events/projects.
    The major event or project could be a marketing program or an engineering upgrade. In either case, an investment is made in inventory because there is only a small window of opportunity in which to use/move the parts and it is considered that the supply chain cannot provide the quantity of parts required in that window of opportunity.

In all three cases, the real driver for investing in inventory is to minimize the risk of other potential costs or losses. Therefore, it is essential to continue to think of inventory as an investment made with the purpose of minimizing risk.

Why Spares and Indirect Inventory Are Different

While the process and actions discussed in this book can be widely applied to many types of inventory, they are particularly effective on spare parts and indirect inventory. Indirect inventory can be defined as the stock of all the "bought in" materials that companies might use. This includes parts and components (in assembly operations), finished goods (for wholesalers), service parts, OEM spares, operating supplies, engineering spares, industrial supplies, and MRO (maintenance, repairs, and operations) parts. Conversely, 'direct inventory' is those materials that a company controls and manages along its own supply chain. This includes work in progress (WIP) and finished goods that are created by the manufacturer.

When most companies (or consultants!) work on improving the management of spares and indirect inventory, they typically fall back on the solutions that are applicable to direct inventory. These solutions are well known and proven for direct inventory. However, when it comes to indirect inventory they do have one fatal flaw — they are applied in a 'one size fits all' approach. This works perfectly well with direct inventory because large quantities follow the same supply path and have reasonably predictable usage, but this is not the case with spares and indirect inventory.

There are in fact six reasons why spares and indirect inventory are different and why the 'one size fits all' approach does not work.

1. The demand is less predictable.
   With most types of inventory, demand forecasting receives a huge amount of attention. With stable and predictable supply chains, an accurate forecast is seen as being the best way to manage inventory. In these cases a forecast that is wrong by 10–20% can cause significant problems. However, with some spares and indirect inventory, the demand could vary by 100% and this would still be acceptable. Consider an engineering spare where the usage is twice what was expected or where there has been no demand, yet the item is kept in stock. Managing this order of magnitude difference in predictability requires a different approach to spares and indirect inventory management.

2. There is usually a large number of Stock Keeping Units (SKUs).
   It is common for organizations that are managing spares and indirect inventory to have thousands or tens of thousands of SKUs. There are few manufacturing organizations that can claim to have that number of SKUs in their direct inventory.

3. The supply characteristics may be different for each and every SKU.
   In addition to the large number of SKUs, another attribute of spares and indirect inventory is that the supply characteristics of each and every SKU may be different. The items will, almost certainly, come from a wide range of individual suppliers and rarely does an organization purchase sufficient quantities of a single item to be able to dictate the logistics of supply.

4. The value and volume of SKUs varies greatly.
   Not only is there a large number of SKUs with different supply characteristics but also the value of individual items and the volume of individual items will vary significantly. With spares and indirect inventory, management must be able to economically order, receive, handle, and store components that cost a few dollars and components that cost thousands of dollars — all with the same system and approach. Some of those components will be supplied in ones or twos and some in the hundreds. Again there is significantly greater variation than with direct inventory.

5. Stockout costs can be disproportionately high.
   When a company runs out of direct inventory, the cost of that stockout is generally limited to the profit margin from the sale that is lost. With spares and indirect inventory, stocking out of a two dollar component may result in thousands of dollars in lost production. This potential for significantly disproportionate costs for a stockout drives many companies to overstock spares and indirect inventory 'just in case.'

6. In some circumstances a low stock turn may be acceptable.
   A high stock turn rate is a goal for most inventory management. This figure can be used as an indicator of the efficiency of inventory management. However, with spares and indirect inventory, it is sometimes acceptable to have very low stock turn rates. In those cases, the low stock turn rate is usually a function of the unpredictable demand, the high stockout costs, and long lead times for supply.

Companies should always seek to maximize their stock turns. But they also need to recognize that an acceptable stock turn for one type of inventory may not be acceptable for another.

The degree to which any one of these issues impacts an organization depends upon the organization's individual circumstance. However, one thing is certain. With this number of issues to differentiate spares and indirect inventory from direct inventory, the solutions used for direct inventory management cannot be applied in a 'one size fits all' fashion to these types of materials.

An Introduction to the Materials and Inventory Management Cycle

Materials and inventory management involves much more than just reviewing the maximum holding level and checking items into and out of a storeroom. Materials and inventory management involves a cycle of activity that starts when the initial need for an item is recognized and then works through setting parameters, procurement/ordering, delivery, storing, issuing, and reordering. This Materials and Inventory Management (MIM) cycle is shown schematically in Figure 2-1.

Notice that each of the steps in Figure 2-1 has an arrow that feeds back into the step in which it originates. This arrow indicates that an internal process exists for that step; it is shown to indicate that at each of these steps the decision making is not a simple one dimensional activity. At each of these steps there are a number of internal processes and even individual behaviors and biases that can and will affect the outcome of that step. In addition there is the internal activity of Return to Store (RTS) that can short circuit the rest of the use–reorder–restock cycle. Although this figure is a simplified representation of the materials and inventory management cycle, it demonstrates that inventory management is anything but simplistic. This point is discussed in greater detail in Chapter 4: People and Processes.

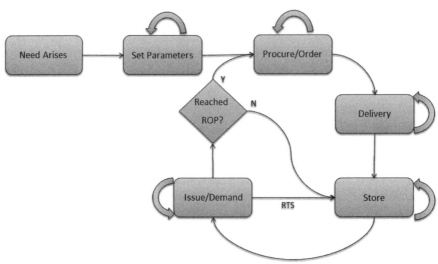

Figure 2-1: The Materials and Inventory Management (MIM) Cycle

Comparison of Theoretical and Actual Inventory Management and Control

Inventory management and control refers to the actions associated with keeping the stock level of a particular SKU within predefined parameters. Figure 2-2 shows a classic 'saw tooth' diagram representing the theoretical movements of an SKU as it is used and reordered. In this diagram, the x-axis represents elapsed time and the y-axis represents the quantity on hand. This figure also shows how some of the definitions mentioned previously relate to the classic saw tooth representation.

The key simplifying attributes of the theoretical model are linear demand (that is, constant and equal demand over time) and instant and complete replenishment. In theory, when demand hits the reorder point (ROP), an order is placed for a predetermined quantity without need for further reference to the users of the item. There is then constant consumption over the lead time while the items are delivered. All items are delivered in one delivery so the item is completely restocked. The theoretical maximum is the Safety Stock level plus the ROQ. In the event that delivery takes longer than expected or there is greater demand than expected during the lead time period, then the quantity on hand dips into the safety stock (which is OK) and the item is completely restocked during subsequent cycles.

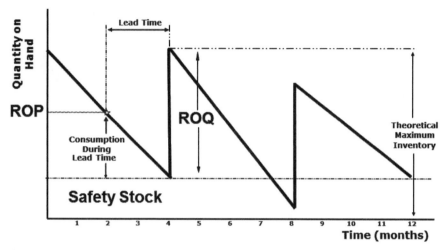

Figure 2-2: The Classic Theoretical Saw Tooth Diagram

The problem is, of course, that reality almost never looks like this. For engineering and spare parts, the chart in Figure 2-3 is far more representative. This graph has four characteristics that separate it from the theoretical profile. These are noted as A, B, C, and D on the chart and explained subsequently.

17

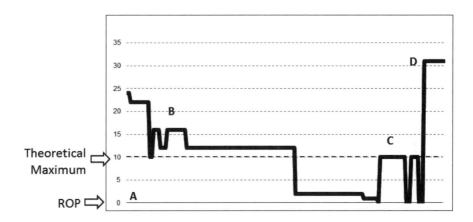

Figure 2-3: Actual Component Demand/Supply Chart

Point A: For this particular component the decision was made to set the initial parameters with an ROP of zero. That is, there is no safety stock. This level is more common for engineering materials and spare parts than many people realize and is not presented here to suggest that this ROP is either right or wrong. It is mentioned because this does not fit the common simplistic theoretical model that insists upon safety stock.

In this specific case, the ROQ was set to 10; hence, the theoretical maximum is 10 (ROQ + Safety Stock). Notice, however, that for the majority of the elapsed time in the chart the actual holdings are much higher than 10. Also, as the holdings rarely reach zero, there is nothing to suggest that setting the ROP at zero is inappropriate. Curiously, there are two instances where the holding increases without having reached zero — this is a clue to what is really going on, which will be discussed shortly. Thus, a traditional review of the ROP and ROQ would provide no improved understanding of how to manage this inventory item because the other elements of the MIM Cycle have far greater impact on the result than just the basic ROP and ROQ settings.

Points B and C: Notice that for this item there are long periods of no movement followed by short periods of multiple movements. Compare this to the theoretical model that assumes a constant and linear usage of items. The difference with the actual profile tells us that the average demand value that is so often used would vary enormously depending upon the point in the timeline at which the snapshot is taken; it is not constant or linear.

It is also interesting to note that the item is expected to be used in sets of 10 (hence the 10–0 setting). Yet of the nine issues of stock within the time line, only three are for the full set of 10. Clearly the management of this item requires insight beyond the obvious idea of setting a simplistic maximum and minimum.

Point D: Now notice the large spike in holdings on the right hand side (at the end of the timeline). This is the real issue with this particular component that was alluded to previously. This spike did not result from additional purchasing, but from a massive and sudden return to store (RTS) of items previously removed. Thus the apparent cycle of usage at point C was not usage at all (although the items were definitely removed from the storeroom). The purchases that were made to replace these items that were not actually necessary. (However, this was not known by those doing the purchasing; they were following their process.) The problem was that the maintenance people who removed the items did not advise anyone that they were not used (or that they may not be used). So, when they eventually had a cleanup and returned the items to the store they were now overstocked, compared to the theoretical maximum, by 21 items or 210%!

This example shows that the theoretical model and the actual situation can be sufficiently different so as to make the application of simplistic solutions not only pointless, but also even dangerous to company finances. A smart inventory solution is to ensure that the influence and complicating factors of all the elements of the MIM Cycle are considered for their impact.

Determining the Re-Order Point (ROP)

Now that some limitations of materials and spares inventory management theory are recognized, we must also acknowledge that someone must at some time determine when to order more stock. Deciding when to reorder requires calculation of the Reorder Point or ROP.

A number of different approaches are used to calculate the ROP, but once again a simplistic approach will not provide the best result. Calculation of the ROP requires consideration of a number of characteristics which help determine the approach that is best for that specific inventory item.

Considering these characteristics is a reality that is missed by many software solutions that use just one approach. (Recall the previous discussion that the word inventory is used a collective noun to describe all the items held, although an inventory is actually made of many separate items that each have their own distinct characteristics).

In determining the ROP, the three main characteristics to consider are the level of demand, frequency of demand, and the probability and impact of a stockout.

Level of Demand

As we saw in the example above, demand is often represented as a perfectly linear equation. However, a linear outcome is more usually not the case. It is the variability in the level of demand that adds complexity to the calculation. This is why forecasting of many inventory types is such a widely-studied discipline. In order to calculate the ROP, you must understand the variability in the demand, not just know the average demand. Here's why.

If the demand for an item is always for the same quantity on each demand event — for example, one electric motor or a set of four spark plugs — then a Poisson distribution is the most appropriate statistical model. (See Figures 2-4 and 2-6 for a summary of different statistical models). Note that at this stage we are considering the quantity, not the frequency, of demand.

If, for any demand event, a variable number of items may be required (for example 3 one time, 2 the next time, 5 the next time), then a Gaussian (or normal) model would be more appropriate. Without understanding both the level and variability in demand, you cannot select the most appropriate method of review.

Frequency of Demand

If the item in question has infrequent demand (sometimes referred to as slow moving), then there will most likely be insufficient data to use a Gaussian model. Again, a Poisson model will be most appropriate. Conversely, high levels of demand will lend themselves to a Gaussian model.

A word of warning: be sure to understand the demand pattern over as long a period as possible. As we saw previously, demand data in a short time frame can be misleading.

Probability and Impact of a Stockout

Strictly speaking the probability and impact of a stockout are two characteristics, but here they are treated as one decision variable because they actually give each other context and are often misused.

The probability/impact decision is often used by practitioners as a reason (or excuse) for overstocking their inventory. The argument that is most often used is that the impact of a stockout is so costly that it overrides any consideration of the cost of the items stocked. This is especially so in industries where the cost of operational downtime is high. However, stocking more than might be needed based on physical limits or probability is pointless and a waste of money. (See also the section in Chapter 5: *When is Critical Really Critical?*) In terms of calculating the ROP, the probability/impact decision affects the service factor component of the calculation. It is, in effect, a risk decision.

Using a Gaussian model, the service factor is a part of the safety stock calculation (see Figure 2-4) and the values can readily be looked up in widely published tables. Figure 2-5 shows a sample calculation of the ROP using a Gaussian model.

Using a Poisson model (Figure 2-6), there is no explicit service factor and the risk element is accounted for in the probability part of the model. Here's how that works. The Poisson function calculates the probability of a certain level of demand over a period of time. If you set that quantity as your ROP, then the probability can be treated as your service factor. Your risk of a stockout is 100 minus the probability of that level of demand.

So, if you look at Figure 2-6, the probability of 7 or fewer demands is 96.4%. Therefore, the risk of a stockout, if you have a reorder point of 7, that is — the risk that there will be more than 7 demands during the lead time for restocking — is:

$100 - 94.9 = 5.1\%$.

The major issue though with the probability/impact decision is the Service Factor Trap. The service factor is the percent of time that the storeroom can supply the required item when it is needed. So a theoretical service factor of, say, 97% sounds high, but in reality for engineering materials and spares, this may not be acceptable.

21

First, if measured across the entire inventory, no one will care about the 97% figure if the 3% includes critical parts and your plant is shut for a week while they get air-freighted in!

Second, you can have a high overall service level and still be significantly overstocked in individual items, meaning that you have still spent money on items that are not needed.

This is the Service Factor Trap. It can be misleading in terms of the inventory being able to fulfill its actual requirements and in terms of how efficiently money has been invested in inventory. Sweeping statements relating to service factors are convenient and reassuring, but add no real value to the practice of materials and inventory management.

The impact characteristic also depends upon where are you located. Consider a situation where a machine will not run without a specific part. Without doubt, this part would be considered critical and the impact of a stockout significant. However, if you are in an urban center with lots of suppliers close by you, may be able to convince one to hold the part for you and then get the required part delivered within an acceptable time frame – for instance while you remove the failed part. However, if you are located in a remote area where delivery takes days, then the stockout has more significant implications. Both situations have the same probability of failure and at one level the same impact — that level is the plant stops. However, the real impact is different if the full materials management cycle is taken into account. The one size fits all solutions that get rolled out to every situation do not bring the required results.

Location, culture, operating mode, financial status, reliability, and risk tolerance are all things that need to be taken into consideration when determining the ROP.

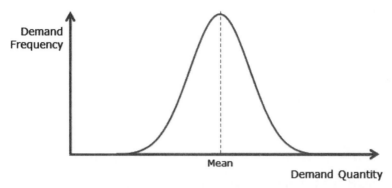

ROP= (Demand x lead time) + safety stock

Where:
    D = Demand in units per week
    LT = Lead Time in weeks
    Safety Stock = csf x MAD x √LT
    csf = Customer Service Factor
    MAD = Mean Average Deviation
Then:

ROP = (D x LT) + (csf x MAD x √LT)

Figure 2-4: The Normal or Gaussian Distribution

Assume:
- Demand = 5/week
- LT = 4 weeks
- MAD = 1 (relatively low volatility)
- csf = 2.56 (assumes 98% availability)

Then:

ROP= (Demand x lead time) + safety stock

ROP = (5 x 4) + 2.56 x 1 X Sqrt(4)

ROP = 20 + 5.12

ROP = 26

Notes:
1. This calculation uses a Mean Average Deviation (MAD) rather than a Standard Deviation. MAD is a simplified way of determining the deviation and is calculated by determining the average value by which demand deviates from the mean, in absolute terms.

2. The Customer Service Factor is based on a MAD scale, not the Standard Deviation of a Normal curve.

Figure 2-5: Example ROP Calculation

Expected number of demands for the item in the supply lead time   4

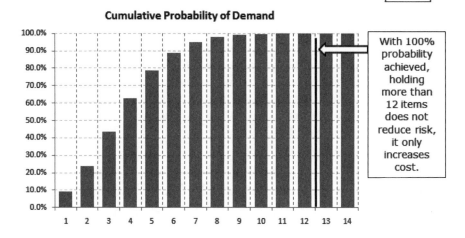

No. of Spares Held	Probability of Event	Cumulative Probability
1	9.2%	9.2%
2	14.7%	23.8%
3	19.5%	43.3%
4	19.5%	62.9%
5	15.6%	78.5%
6	10.4%	88.9%
7	6.0%	94.9%
8	3.0%	97.9%
9	1.3%	99.2%
10	0.5%	99.7%
11	0.2%	99.9%
12	0.1%	100.0%
13	0.0%	100.0%
14	0.0%	100.0%

The Probability of Event is the probability of that exact number of demand during the lead time.

The Cumulative Probability is the probability of that number or fewer demands during the lead time.

In this example, there is a 6.0% probability of exactly 7 demands and a 94.9% probability of 7 or fewer demands during the lead time for restocking this item.

Figure 2-6: The Poisson Distribution

Determining the Re-Order Quantity (ROQ)

The other key decision for materials and spares inventory management is to calculate the Reorder Quantity (ROQ). The ROQ is usually not so highly discussed as the ROP but it has as much, if not more, impact on the quantity of inventory that is held. This is because the point at which a company actually commits to holding inventory and tying up working capital is when the items are ordered. The classic formula for calculating the 'economic' ROQ is as follows:

$$ROQ = \sqrt{\frac{(2 \times \text{Order Cost} \times \text{Demand Rate})}{(\text{Item Cost} \times \text{Holding Cost P.A})}}$$

Where:

Order Cost = the company internal cost for processing requisitions, issuing purchase orders and receiving deliveries.

Demand Rate = the expected demand over a year.

Item Cost = the purchase cost of the item, including delivery costs.

Holding Cost = the financial charge for holding inventory (see Chapter 3: The Financial Impact of Inventory).

Although simple in concept, there are some complications in practice.

1. The order cost is crucial to the calculation.
   Of the four variables in the calculation, this is the least simple to determine because there is no set rate. The actual order cost will be different for every company and is dependent upon internal company efficiency, local pay rates, and so on. To calculate the order cost, some companies use an Activity Based Costing approach; some just use an estimate such as $100 per order. Note that it is a mistake to use a simplistic calculation such as the total cost of the purchasing and stores departments divided by the number of orders placed because this assumes 100% capacity utilization. No matter which approach you choose, the key is to understand the impact of an error in this

value. From the formula you can see that the ROQ varies directly with the square root of the order cost. So, if your estimate of order cost is two times the actual order cost, you will be ordering 41% too much stock and that could be a lot of money. (Recall that the square root of 2 is 1.41.) This estimation error is simple to make. Let's say the real order cost is $50 per order, but you decide to use $100, just to be sure that everything is covered. This doesn't seem like much, but will add 41% to the quantity of inventory purchased.

2. The formula assumes that the order cost is fixed.
   Your actual order cost may vary due to efficiencies related to the supplier. This could include extra costs at your end due to the supplier being inefficient, losing paperwork, hard to contact, requiring follow up, and order expediting. Or the costs could be less due to acceptance of blanket orders, use of electronic methods, and so on. A blanket approach could result in significant overstocking and the calculated ROQ should be reviewed for any orders of a significant value.

3. The formula assumes that the demand is constant.
   We have already seen an example where the demand varies significantly over time. If the calculation is performed when the demand is high, the calculated ROQ value will be high and you will be overstocked.

4. The formula assumes one delivery per order, no allowance for scheduling or batching.
   Not all orders are delivered in one delivery and each delivery costs you money in terms of workload.

When faced with determining the factors used to calculate the ROQ, it is suggested that values are used that minimize the order quantity rather than maximize the order quantity as this is the lesser of two evils.

If you overestimate, you will spend too much on stock and unnecessarily tie up money or, worse, spend money on items that might never be used. This type of error is rarely addressed because it does not automatically trigger any action. However, if you underestimate your ROQ — and assuming that your ROP is appropriately set — then you will only end up ordering more frequently and this can trigger the need for a review.

You can then set the ROQ at a more appropriate level. The effect of different order costs is shown below.

$$ROQ = \sqrt{\frac{(2 \times \text{Order Cost} \times \text{Demand Rate})}{(\text{Item Cost} \times \text{Holding Cost P.A})}}$$

1. Assume that:
    Order Cost = $100
    Demand = 1,000 per year
    Item Cost = $10
    Holding Cost = 25%

    $$ROQ = \sqrt{\frac{2 \times \$100 \times 1,000}{\$10 \times 0.25}}$$

    $$= 283$$

    Therefore, the 'economic' ROQ is 283 items.
    This means that this item will be ordered, on average, 3.5 times per year (1,000 per year/ 283 per order).

2. Let's look at the impact of changing the Order Cost. Assume that:
    Order Cost = $50
    Demand = 1,000 per year
    Item Cost = $10
    Holding Cost = 25%

    $$ROQ = \sqrt{\frac{2 \times \$50 \times 1,000}{\$10 \times 0.25}}$$

    $$= 200$$

Therefore, if the order cost is really $50 per order the 'economic' ROQ is only 200 items – approximately 30% lower than if the order cost is $100. This means that this item will be ordered five times per year (1,000 per year/ 200 per order).

A Word on Monte Carlo Simulation

Monte Carlo simulation is a complex analytical technique that uses random numbers as input variables and applies them to a known function (or formula). It is reportedly named after the random inputs that occur in table games, such as roulette, at the casinos in Monte Carlo.

With inventory analysis, it removes the constraint of having to make assumptions about the frequency or level of demand as these would be randomly generated values. When used in a computerized simulation, the technique can run through a high number of cycles to demonstrate under which scenarios supply would not be available. From this perspective, it appears to be an attractive option for inventory review and is widely used in the academic analysis of inventory management.

The technique does, however, suffer from the same shortfall in practice that limits most analytical approaches — it does not easily enable consideration of the entire materials and spares inventory management process. Instead, it focuses solely on the mathematical evaluation of the ROP and ROQ settings.

Do You Hold Too Much Inventory?
Check Your Stock Turn Ratio

There are a number of measures that get used for tracking inventory performance. One of the most popular measures is stockouts. A stockout occurs when there is demand for an inventory item but there is no stock.

It is essential to measure the availability of stock. After all, that is why the investment is made in the first place. However, measuring stockouts can be a limiting way to measure inventory as it only measures one dimension of inventory, that is, availability. This approach is limiting because one way to ensure a low number of stockouts is to overinvest in inventory so that stock is always available no matter what. This is sometimes referred to as 'just in case' inventory.

What Is a 'Stock Turn'?

Because inventory requires a significant financial investment and that investment involves significant ongoing costs, it is also important to measure the financial performance. Tracking the value of inventory is important for cash management purposes. However, an additional financial measure that often gets overlooked is the stock turn ratio.

The stock turn is calculated by dividing the annual usage of the inventory (in dollars) by the value of the inventory held (also in dollars).

$$\text{Stock Turn} = \frac{\text{Annual Usage}}{\text{Value Held}}$$

For example, if a company holds $5M worth of inventory and issues $2.5M worth of that inventory in a year, the stock turn ratio is 2.5/5.0 = 0.5. That is, the company turns over its inventory at the rate of one half per year. Obviously, the higher the stock turn ratio, the better.

What Stock Turns Tell You

Stock turns measures the efficiency of the inventory investment by telling you whether you have overinvested in inventory and whether you have the right mix of inventory. (Note, however, that it won't tell you about specific inventory items.) For example, if the number of stockouts is low (which is good) and the stock turn ratio is also low (which is bad), you have an indicator that there may be an overinvestment in inventory. If the number of stockouts is high (which is bad) and the stock turn ratio is low (which is also bad), then you may have invested in the wrong inventory. That is, your money is tied up in stock that doesn't turn over and you hold too little of the stock that is in demand.

Stock Turn Targets

An appropriate target for stock turns in your business will be influenced by a range of issues, some within your control and others outside of your control. For example, if you have spares that are imported from somewhere far away or you are in a remote and isolated area, then you are likely to hold more safety stock and, therefore, have a lower stock turn.

Conversely, if you are located in a densely-populated area surrounded by similar industry and many suppliers, you should be able to achieve a high stock turn. But this isn't the whole story because if your processes don't adequately control decision making on materials and spares inventory stocking, you are also likely have a low stock turn.

Using Stock Turns as a Key Measure

The key thing to remember when using a stock turn ratio is that it must be applied across the entire inventory. You cannot 'cherry pick' elements of inventory. The reason for this is that some inventory items will naturally have a high turnover and some will be low. The aim of the ratio is to measure the overall efficiency of the inventory investment.

In one recent case, an inventory manager tried to justify the size of his inventory by pointing out that one section of inventory had a stock turn of 5 (very good in his circumstance) and that another section had a stock turn of 0.2 (very bad). The justification was that insurance spares caused the low stock turn and, therefore, nothing further could be done. This analysis, however, ignored a large component of inventory that could be managed down and it ignored the possibility of consignment stock for the fast movers.

As mentioned above, stock turns is also a great measure to use when you have multiple sites or locations within the one company. As an internal benchmark, stock turns readily shows which sites have better control over their inventory.

Stock turns is an essential measure of inventory performance because it measures the inventory efficiency. When used in conjunction with other measures such as stockouts, the overall performance of your inventory investment can be determined.

Chapter 3

The Financial Impact of Inventory

Don't Overcompensate

How often have you heard someone say, 'Without this inventory, production will stop' or 'We are merchandisers; we need our inventory or we won't make sales.' These statements represent the typical view of inventory—more is better! It is true that without the right inventory some sales may be missed or a production line may stop. Both are outcomes that may result in lost revenue. It is also true that too much inventory costs a business even if it is the right type of inventory. It costs money to buy inventory, it costs money to store inventory, and it costs money to finance inventory.

Often people say that the real issue is one of balance—that is, balancing the cost of inventory investment with the potential gain (or loss) from not making the investment. However, this view is overly simplistic. It can be misleading if applied universally. The goal should be to ensure that you don't overcompensate for uncertainty by stocking materials that you just won't need and to avoid overinvesting in the inventory that you do need.

In many cases, the ongoing level of investment that companies make in their inventory just does not make sense. These are the cases where stock minimums are never reached. Perhaps months of supply are held when weeks (or days) will do. Or the inventory is just not needed, but is not sold off or otherwise removed from the system. These are the cases where the ongoing cost of the inventory just cannot be justified.

The Inventory Management Tension

The conflicting needs of meeting demand expectations and minimizing the financial investment in inventory sets up a tension between Operations and Financial Management. This inventory tension is shown in Figure 3-1.

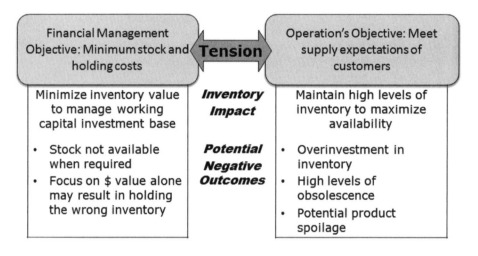

Figure 3-1: The Traditional Inventory Tension

Financial managers such as Chief Financial Officers (CFOs) and accountants will typically seek to minimize the working capital that is tied up in a business. Working capital includes inventory. But if these managers take an approach that is based solely on the dollars invested, it is likely that the investment will end up being made in the wrong inventory. Why? Because the aim is to reduce the dollars invested, not to meet demand expectations.

The accountant's approach will result in inventory being cut in any way possible to drive a working capital outcome. This approach is most likely going to be the case if the performance of the financial manager is measured solely by financial outcomes such as working capital, not operational or sales outcomes.

Conversely, operations managers typically seek to maintain a high level of investment in inventory. Their goal is to ensure that inventory will be available to meet demand. As a result, there is likely to be an overinvestment in inventory; there may be high levels of obsolescence, and, in some cases, high product spoilage. These outcomes occur because operations managers are typically measured by plant performance or Profit and Loss metrics (P&L — sometimes referred to as the Statement of Financial Performance) such as revenue, cost, and EBIT (Earnings Before Interest and Tax). Inventory is a balance sheet item (see below) and does not impact the P&L. Therefore, minimizing inventory is typically of little concern to operations managers.

Resolving this inventory tension requires an approach to materials and spares inventory management that strips away the excess and ensures that a company only stocks the right amount of the right inventory. The first step to achieving this is to understand the financial aspects of materials and inventory management.

Cash Is King

There is an old saying in business that 'Cash is King.' What this means is that cash is the lifeblood of all business. No cash, no business. Spending money on materials that end up in inventory ties up cash and diverts it from other potentially revenue-generating or cost-saving uses. The aim, therefore, should be to minimize the inventory investment for a particular level of customer service. The approach taken should ensure that the target level of service is met while also minimizing the cash investment. In turn, this approach will maximize the overall benefit for the company.

A Simple Model of Business Economics

Understanding the importance of cash requires having a basic understanding of business economics. To help explain this, turn to Figure 3-2, which represents a simple cash flow cycle.

(1) Starting at (1), the 12 o'clock position, this business has some 'Cash on Hand' with which to operate. The cash might have come from investors or it might have been borrowed from the bank.

(2) In either case, the investors or the bank will want a return for making the money available. The return will be either interest for the bank or dividends for the investor. In a real business, some of this cash would be used for investment in plant, equipment, and buildings. But for simplicity, in this example we will assume that that type of investment is complete and requires no further funding.

(3) The business spends its cash on buying raw materials, spares, labor, and utilities so that it can make products.

(4) Typically the business will need to hold inventory of the raw materials and spares in order to be able to provide a buffer between supply of these materials and the demand from their production department. The buffer is necessary because the rate of demand is usually greater than the ability to supply. In order to accumulate inventory for this buffer, they need to buy more than they will actually need in the short term.

(5) The raw materials are then used to make product. After the product is made, it will be put into inventory, again to act as a buffer between supply and demand. Again, to accumulate the inventory for the buffer, the business must make more than is needed in the short term.

(6) When these products sell, the buyer pays the company; the payment provides an input, or receipt, of cash to enable the company to recommence the cycle.

While the cash flow cycle continues, the company will need to spend cash on overheads, new investment, interest on money borrowed, repayment of money borrowed, and dividends to shareholders.

A company's cash flow is the difference between all the cash that goes out (buying raw materials, spares, utilities, labor, overheads, investment, dividends, interest, and loan payments) and the money that comes in (receipts from customers).

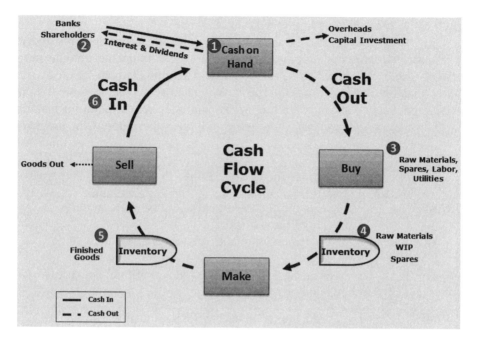

Figure 3-2: The Cash Flow Cycle

If the cash flow is negative, that is, the 'cash in' is less than the 'cash out', the company will need to borrow more money or it will be unable to buy supplies, labor, utilities, etc. As mentioned previously, no cash, no business. If the 'cash in' is greater than the 'cash out' then the cash flow is said to be positive. The business will have excess cash it can then invest or use to repay borrowings.

Here is where materials and inventory management comes in: if a company is able to reduce its level of inventory, it can free up the cash that is invested in the buffers between suppliers, production, and customers. This improvement then reduces the 'cash out' and helps provide a positive cash flow. The extra cash can then be used to pay back borrowings and reduce interest bills or it can be used for further capital investment, without borrowing more money. This is the most effective way for a company to manage its cash — free up the cash that it has already invested in itself.

Minimizing the investment in inventory is good business practice for all companies as they improve cash flow. It produces both an increase in cash and an ongoing reduction in costs. Table 3-1 shows how reducing inventory has a positive impact on the key business measures of cash flow, ROFE (Return on Funds Employed), and EBIT.

There are a number of actions that can be taken to reduce inventory; these are discussed in detail in Chapter 8. However, to demonstrate the cash effect, Table 3-2 shows some of the generic actions that can be taken and their specific business impact. The key point to remember is that 'cash is king' and that excessive inventories are an unproductive use of cash.

Business Measure	Impact of Reduced Inventory
Cash Flow	• Reduces the outflow of cash because items are not purchased for restocking • Delays expenditure until an item is more likely to be needed • Frees up cash for other uses such as capital upgrades or repaying borrowings
ROFE	• By reducing the investment base, the return on funds employed (ROFE) increases • By increasing ROFE, the company becomes a more attractive investment • The impact of this change is likely to be an increase in share price
EBIT	• By freeing up cash to pay back borrowings and/or minimizing the amount of borrowings, the company saves on interest payments and holding costs. This adds directly to the company's earnings

Table 3-1: The Impact of Reduced Inventory

Example Actions	Impact
Remove obsolete stock	• Scrapping stock will provide a tax benefit in most countries. • Sale of items will provide an inflow of cash (and may produce a profit).
Reduce reorder stock	• A delay in spending cash retains cash in the business (improving cash flow). • Lower reorder quantities reduce the stock held value and associated costs.
Reduce maximum holdings	• Reduces the stock held value and associated holding costs.
Remove overstocked items	• If removed by natural attrition, the impact is to reduce the stock held value and associated costs. • If written off, the impact is the difference between held value and revenue if sold.
Reduce quantity held	• In addition to the obvious reduction in inventory, the impact will be to reduce the costs of counting, maintaining, moving, storing, etc.

Table 3-2: Examples of Actions and Their Impact

The Four Financial Reports that You Absolutely Must Understand

Now that you understand the business impact that inventory has on an organization, it is important to understand the financial reporting that relates to inventory. There are four levels of reporting that you absolutely must understand:

1. The Cash Flow Statement
2. The Balance Sheet
3. The Profit and Loss Statement
4. The Operating Statement

Each of these reports relates to financial outcomes. However, the type of expenditure and the time frame that they address vary. Understanding these distinctions is crucial to appropriate materials and spares inventory management.

1. The Cash Flow Statement
As described in the Cash Flow Cycle (see Figure 3-2), businesses have cash that comes in (capital raising, loans, revenue from sales, etc.) and cash that goes out (capital purchases, payments to suppliers, wages and salaries, dividends, etc.). The Cash Flow Statement is simply an accounting of each of these inputs and outputs so that we can readily see where money came from and where it was spent. The Cash Flow Statement typically classifies the transactions in cash flows relating to operating activities, financing activities, and investment activities. Money that is spent on materials and spares will appear on the cash flow statement as an outflow of cash in the 'operating activities' section under a title such as Payments to Suppliers.

2. The Balance Sheet
The Balance Sheet is sometimes now referred to as the Statement of Financial Position because it is a snapshot of the financial assets and liabilities of a company. As a snapshot, it describes the state of the business on the day that the report represents. The content of the balance sheet, or at least the items represented and their meaning, is

standardized through international accounting standards that form the legal reporting requirements for all companies (although the requirements may vary slightly from country to country).

Materials purchased as inventory will appear on the Balance Sheet as an asset because, in theory, they have a financial value (see Figure 3-3). They could be sold to raise cash (which of course is what happens with raw materials, WIP, and finished goods inventory). When the inventory materials are purchased, their cost is recorded against a 'capital account' which is like a cost center or cost code for balance sheet items.

Current Assets	$M	
Cash	30.0	
Receivables	60.0	Materials and
Inventories	25.0	spares will
Other	10.0	be included in this entry
Total Current Assets	125.0	
Non-current Assets		
Receivables	2.0	
Inventories	5.0	
Property Plant and Equipment	500.0	
Intangibles	5.0	
Other	10.0	
Total Non-current Assets	522.0	
TOTAL ASSETS	647.0	

Figure 3-3: Example of the Asset Section of a Balance Sheet

Although companies can generate a balance sheet at any time, they are typically produced (or published) twice a year: the fiscal midyear and fiscal year end reporting. From this we can see that, if the inventory on the balance sheet is only reported twice a year, it may only get highlighted and receive attention twice a year. Is it any wonder then that it becomes the forgotten investment! When accountants comment that a company holds too much inventory, they are typically referring to the value on the balance sheet rather than the physical holdings of any specific item.

3. The Profit and Loss Statement

The Profit & Loss (P&L) Statement is sometimes now referred to as the Statement of Financial Performance. Although the Balance Sheet shows the state of affairs at a particular point in time, the P&L compares income with expenses over a period of time. This could be a month, quarter, or year.

The P&L does not include Balance Sheet items such as property, plant and equipment, and other assets, such as inventory. This is because the money spent on capital items and inventory is expected to be consumed over several (or many) reporting periods, rather than in one period. The attribute that it is not used in the same period in which it is purchased is an important factor for identifying inventory. Thus, the P&L does not record the level of inventory held or, overtly, the change in the level of inventory.

Because most operational personnel are concerned with operational expenses and generating company profit, they may be familiar with the P&L but many do not realize that it does not help with inventory management.

4. The Operating Statement

The typical operating statement details only the expenses over a period of time. This statement might also be referred to as a Cost Center Report, Expenditure Report, or Departmental Report. The Operating Statement does not include revenue; it only includes costs and might be issued weekly, monthly, quarterly, or annually (or all four!). The statement includes items such as labor costs, overtime, and material costs, by department.

Engineering materials and spares are only included if their purchase cost is recorded against one of the cost centers included in the report.

This means that they are purchased directly against an operating cost center or are 'booked out' or 'issued' from the storeroom against that cost center. This action moves the cost of the item from the capital account on the balance sheet to the P&L and Operating Statement.

How Much Cash Can You Generate?

Now let's try and put some numbers together to get an understanding of the magnitude of the benefits that you might realize by implementing an appropriate inventory review. This is important for two reasons. First, goal setting is an important element of any improvement program. An appropriate goal provides direction and motivation—a 'yard stick' for measuring progress. Second, goal setting can assist in justifying an investment in the resources required to achieve the benefits. There are few things more frustrating than seeing value-adding opportunities and not being able to justify the investment required to achieve them because the benefits are difficult to quantify.

On page 47 there is a calculation sheet for you to use to calculate the benefits you could realize from a well-implemented program. However, before you complete that form, work through the example on page 46; it will help explain the elements of the calculation. Turning to Figure 3-4 you will see that the sheet is divided into two sections, Inventory Costs and Potential Savings.

Step 1: Let's first look at the Inventory Costs. The topmost number on Figure 3-4 is the current inventory value. In this example it is $5,000,000. For your organization, this number might be more or it might be less. The important thing is that you record the value as reported by your Finance Department—don't guess.

Step 2: The next line down reads Estimated WACC. The cost of financing working capital is much more than just the interest rate that businesses pay on borrowings. In business finance, there is a term called the Weighted Average Cost of Capital (or WACC, pronounced wack). The WACC will be different for every company because it is based on the sources of capital, such as shareholder funds, borrowings from banks, bond issues, and so on. Because this book isn't a business finance text, I won't be explaining this concept further except to say that for most companies the WACC generally ranges from 10–15%. To be conservative, the calculation in the example will use 10%. You should ask your Chief Financial Officer (CFO) what WACC to use for your company.

Step 3: The next line is for estimating the value of obsolescence and spoilage as well as the costs of managing and storing your inventory, as a percentage of the total inventory value. It is fair to say that this is usually the most difficult number to estimate. Some companies estimate that this cost could be as much as 25% of the total investment in inventory per year. However, for general use, a value of 10% is recommended as a suitable rule of thumb.

Step 4: Now by adding together the values from Steps 2 and 3, you get what I call the Inventory Cost Ratio. This ratio represents the annual percentage cost for just having the inventory available. This does not include the cost of actually purchasing the materials. In this example, the Inventory Cost Ratio is 20%.

Step 5: Multiply the Value of Inventory by the Inventory Cost Ratio to determine the Total Annual Cost of Inventory. In this case, $5,000,000 x 20% = $1,000,000. This is the annual dollar cost just for having the inventory available. It does not include the cost of actually purchasing the materials. Instead, this is how much it costs simply to hold this inventory each and every year.

Now calculate the benefits of an inventory reduction program.

Step 6: To calculate the benefit, the first thing to do is to estimate a target. Elsewhere in this book you will find case studies where companies have achieved results as high as nearly a 50% reduction. However, in my experience, the average is somewhere around 25%, so let's use that value.

Step 7: To calculate how much cash you can potentially realize, multiply the Inventory Value by the Reduction Target. In this example, this is $5,000,000 x 25% = $1,250,000 and is shown as the Potential Cash Release. This calculation indicates that a program of inventory reduction with these values could realize a cash benefit of $1,250,000. That's a lot of capital!

Step 8: In addition to the cash savings, you can also generate an ongoing saving because the cash generated in Step 7 no longer needs financing or will result in obsolescence. This means that you generate annual savings equal to the Potential Cash Release multiplied by the Inventory Cost Ratio. In this example, this is $1,250,000 x 20% = $250,000. This amount is in addition to the Potential Cash Release and will be saved each and every year.

Inventory Costs

Current Inventory Value	$5,000,000
Estimated WACC	10.0 %
Estimated % Cost of Obsolescence, Spoilage, Managing, and Storing Inventory	10.0 %
Inventory Cost Ratio	20.0 %
Total Annual Cost of Inventory	$ 1,000,000

This is how much it costs to hold this inventory
each and every year.

Potential Cash Release and Ongoing Savings

Reduction Target	25 %
Potential Cash Release Inventory Value x Reduction Target=	$ 1,250,000
Potential Ongoing Savings Cash Release x Inventory Cost Ratio=	$ 250,000
	Per year

Figure 3-4: Example of a Calculation Sheet

Basic Inventory Data

Current Inventory Value $

Estimated WACC   %

Estimated % Cost of Obsolescence, Spoilage,
Managing, and Storing Inventory   %

Inventory Cost Ratio   %

Total Annual Cost of Inventory $

This is how much it costs to hold this inventory
each and every year.

Potential Cash Release and Ongoing Savings

Reduction Target   %

Potential Cash Release
          Inventory Value x Reduction Target= $

Potential Ongoing Savings
          Cash Release x Inventory Cost Ratio = $
                                        Per year

Figure 3-5 Blank Calculation Sheet

Chapter 4

People and Processes

The People Factor

Ask most people which three factors have the greatest impact on their inventory holdings and they will most likely respond by identifying factors that are largely independent of people and processes, such as:

- Usage
- Risk
- Cost of a stockout

The reality, however, is that most inventory issues are caused by people and management processes. The three factors that have the greatest impact on inventory holdings really are:

1. People
2. Management systems
3. Accountabilities

Management systems and accountabilities are dealt with later in this chapter; this section focuses on the people factor.

The Importance of People

It seems self evident to say that people are the backbone of any materials and spares inventory management system. Sure there may be software and computer systems, defined processes and procedures, corporate policies, geographic constraints, equipment types, supplier terms, and many other influences, but the one thing that ties all these together is people.

It is your people who interpret situations, make decisions, and execute plans. It is your people who don't complete the paperwork, delay ordering, over order, take excess quantities from the storeroom only to return them later, hide spares in their own squirrel stores, decide to hold 'just in case' inventory, make forecasts, fail to plan, and forget to ring, write, or email. It is your people who set conflicting priorities, set measures and targets that influence behaviors, ask for meaningless reports, and fail to hold others accountable for their actions. It is beyond all reasonable doubt that people have the single most important influence on inventory holdings. Yet this is the factor that gets the least attention. Why? Because people are complex. Sometimes they hold perceptions that not everyone understands, and sometimes they are motivated by things that are not always clear.

However, people are also the key to success. Once your people understand <u>why</u> something is important and <u>how</u> their decisions and actions influence the outcome, they then start to take actions that are aligned with the inventory goals. Let me tell you about Brian.

Quite a number of years ago I used to work with a guy named Brian (although for obvious reasons that's not his real name). Brian was the materials manager at a company at which I was the maintenance engineer.

Brian was a terrific guy. Everyone liked Brian. He always seemed to know what was going on. People used to say that he had his 'finger on the pulse.' In fact, whenever we had a problem with shortage of inventory, late deliveries, or 'missing' items that were tucked up in someone's 'squirrel store,' Brian could find it or fix it. Brian took pride in his work and was known to remark occasionally on how much money he and his team had saved the company by the way they responded to the various emergencies that cropped up.

Then one day Brian announced that he was leaving. His work had been noticed and he had been 'head hunted' into a more senior role at a bigger organization with a substantial pay rise. Everyone was sad to see Brian go and we all knew that Brian's replacement had, as they say, big shoes to fill.

Brian's replacement (let's just call him the New Guy) started soon after Brian left and something unexpected happened — the number of emergencies went down and became the rare exceptions rather than the norm.

Of course this didn't happen straight away. In the beginning, things were worse than ever. It seemed that Brian didn't just keep his finger on the pulse; he kept a tight rein on his team. None of them would (or perhaps even could) do anything without Brian's input. Brian made himself the center of everything.

When the New Guy started, he didn't want this so he worked with the team to set guidelines for decision making and new policies and procedures. They could then operate with some greater autonomy. He wanted them to know both what to do and what was expected of them. It is an overused term these days, but the New Guy empowered his team.

The result of this was improved materials and inventory management and fewer emergencies. Plus, there was another interesting side effect — the accident rate in the storeroom and warehouse halved within six months!

So what was going on? Was this just about policy and procedures? Of course not. The real issue was a combination of people, ego, and adrenaline.

Brian was an adrenaline junkie. He received an adrenaline boost every time there was an emergency. In turn, these emergencies had the great ego payoff of him being the one that solved the problem. The instant gratification of being reactive had kept Brian from being proactive and working out how to prevent problems that arose, rather than just solving them.

This is why being proactive is so hard to implement. Being proactive eliminates the immediate feel-good factor and adrenaline rush. In a well set-up system, there are fewer emergencies (I would never say none!). Furthermore, there is no instant gratification for solving the problem. In a well-managed system, you get the warm glow of a job well done not, the instant rush of adrenaline. In a well-managed system, people know what to do and what is expected of them.

Although Brian and I worked together many years ago, I still see many people like Brian in the work that I do with companies all around the world. Where do you stand in your operations — are you like Brian?

It's Really All About Culture

Culture can be defined as: the prevailing behaviors that determine how a company's people respond in a particular circumstance.

Culture can be local, as in applying to the team in a single storeroom. It can also be global in that everyone across a business responds in the same way. Some companies have a safety culture, some a customer service culture, some a 'can-do' culture, and some a 'see-what-you-can-get-away-with' culture. With materials and engineering spares, you need a 'let's-not-spend-more-than-is-absolutely-necessary' culture.

Recently I reviewed the materials and engineering management for an organization that had five major sites with independent storerooms. This organization had company-wide policies and procedures; it consistently used a well-known enterprise resource planning (ERP) software system. In theory, the differences in their inventory holdings should have been limited to local conditions. For four of the five locations, this was the case. However, one location was not just a little better, but was far superior to the others. We could only identify one reason for this — the site manager considered it important to truly optimize their inventory holdings.

This attitude was fostered throughout everyone involved in materials and inventory decision making at that site. As a result, everyone responded and acted in a way that was consistent with the 'let's-not-spend-more-than-is-absolutely-necessary' culture.

People Are Biased

One of the reasons that people sometimes make decisions that are not consistent with expectations or obvious to everyone is that they are biased. Research has shown that there are seven biases that people exhibit with investment decision making (such as buying stocks and shares) and it is worth reviewing these with respect to materials and spares inventory-related decision making.

The seven biases are:

- Overconfidence Bias — this is when you overrate your own skill and ability in a particular area. Problems arise because you don't know what you don't know, leading to substandard decision making.

- Loss Aversion Bias — we all dislike losses, but people most dislike losing face — that is, appearing to be wrong. Operationally this is exhibited when we continue with a strategy or action even though it is not working out. What we say is 'it is just about to pay off, just give it more time.'

- Confirmation Bias — this occurs when a positive outcome is seen to reinforce the appropriate nature of some activity, even if the outcome was not caused by that activity. It may have been caused by something else or just be coincidental.

- Framing Bias — the response to addressing an operational issue is often guided by the options that appear to exist rather than the range of options that are available. This often happens unconsciously because we assume that the options we see represent the range of solutions available.

- Anchoring Bias — this is where you focus on a specific outcome to use as a comparison for success. This bias occurs a lot with benchmarking where a number is determined and is then used as a reason not to take any further action because 'we are near the benchmark.' As a result, many genuine opportunities can be left unrealized.

- Status Quo Bias — related somewhat to the Loss Aversion bias, we often prefer to maintain the status quo rather than risk some potential loss (say of time) by changing to something new.

- Myopic Loss Aversion Bias — taking a short term view of the world can limit real success in the long term.

Chapter 12 includes a case study where these biases were clearly present in materials and inventory management.

To overcome these biases there are several things you can do.

First, be aware of the biases and when you may be falling into the trap of one or another. Next, be rational and not reactive in your decision making; recognize any assumptions and test those assumptions against facts. Finally, reflect on your decisions. The carpenter's rule is to 'measure twice and cut once.' If you revisit your decisions before acting, you may uncover new information or recognize other prior experience. And if you really want to reflect on your biases, seek external advice. Sometimes an independent third party can identify issues with assumptions and decision making that we just don't see ourselves because we are too close.

Adopting a Zero Inventory Mindset

In an ideal world, there would be instant replenishment and zero inventory. Instant replenishment would mean that whenever an item is needed it would be instantly available. Companies would hold no stock because they could get delivery in the required quantity, in an acceptable time frame, all the time.

But you don't live in an ideal world and you can't get instant replenishment. What you can do, however, is take an approach to inventory that questions the need for the inventory or the investment. Adopting a zero inventory mindset is not an action; it is a perspective or framework for your entire future decision making. The key is to ask three questions before making any commitment to hold or add to inventory:

1. Question whether the stock is really needed.
   What are the alternatives to actually holding the stock? Could it be supplied on sufficiently short notice or substituted with another item? Would the user be prepared to wait for delivery without the wait impacting your competitive position? For engineering spares, is there an alternative processing path that could be used while the spare is delivered? Do not forget that all inventory costs money. By questioning the real necessity of adding an item to inventory, you are questioning whether an investment should be made in that inventory.

   This is the point at which most inventory thinking stops—do we need it? Yes. Then go and order some. But good materials and spares inventory management goes beyond this level of thinking and asks further questions before deciding how much to invest.

2. Question who should make the investment.
   Once it is determined that access to an item is required with no notice, it almost invariably follows that people assume that their company should be the one to make the investment. A far better alternative would be to have someone else make the investment. From a cash perspective, having someone else make the investment means that you delay any payment until actually needing the item. Many suppliers are willing to make the investment as it guarantees that they maintain continuing business with your company. Before committing to an

inventory investment, you should always explore the possibilities of consignment stock.

3. Determine how planning and process redesign can minimize the investment.

   If it is determined that an item is needed <u>and</u> that the supplier is not prepared to support their ongoing relationship with your business through consignment, then you need to determine what actions you can take to minimize the investment in inventory. There are many actions that can be taken to minimize the investment and these are detailed in Chapter 8: The 7 Actions for Inventory Reduction.

It is by asking and answering these questions every time an inventory decision is made that you start to shape a zero inventory mindset. Ultimately a zero inventory mindset is not about achieving zero inventories, but about not accepting that you must hold or invest in excess inventory.

A zero inventory mindset is about exploring the opportunities for reducing the inventory investment rather than accepting that inventory is a given. A good parallel is with the approach that many companies take these days with safety. Many companies adopt safety policies such as 'no injuries to anyone ever.' Do they believe that they will prevent all possible accidents? Of course not. Will they do everything they can to prevent accidents and injuries? Yes. It is the same with inventory. Will you need to hold some inventory? Most probably. Should you do everything possible to minimize that investment? Of course!

The Influence of Process

In professional sport these days it is not unusual to hear players and coaches talking about process. They talk about focusing on the process and following the process. Rarely do they talk about scoring a goal, a touchdown, a home run, a point, or achieving a good shot. It's all about process. So what do they mean by this? What they mean by focusing on the process is that they focus on the actions they need to take in order to achieve their desired result. They don't focus on the result itself. The reasoning here is that if you follow the steps required, then the result will look after itself. This is one of the big differences between professional and amateur sportspeople. Amateurs often focus on the result and forget about doing all the things that would almost automatically lead to the result.

Let me use golf as an example. When golfers step up to hit a golf shot, they know that, generally speaking, the closer to the hole, the better the result. They also know that the faster the club head is moving when they hit the ball, the further the ball will fly. This is where the professional and amateur often take a different approach. Amateurs will try to hit the ball as hard as possible in order to go as far as possible. They focus on the result of hitting the ball a long way. When they do this, they often mis-hit the ball or lose control. Professionals don't think about hitting the ball as far as possible; they think about getting the process right. They think about the way they stand, the swing, the rhythm, and their own routine. Professionals don't seek to hit the ball hard; they try to hit it correctly. The irony here is that, as a result, they hit the ball a long way and with great control. So why doesn't every golfer just do this? Good question. The reason, I believe, is that no matter what the amateurs do, they get a result. The ball moves closer to the hole — maybe not as close as it could be or in as good a position, but a result none the less. And most amateurs are happy with this, which is precisely why they are amateurs!

What does this have to do with inventory?

Well, with inventory management the same thinking applies. There is a process and there is a result. Some people follow the process and get great results. Others try to shortcut the process, they focus on the result and they get something less than great results. More important, however, they lose the opportunity to deliver a great result. Often they don't even realize that they have underachieved because the results are not as clear cut as they are on the golf course! To understand the influence of process, we need to understand systems thinking.

Systems Thinking

Systems Thinking was first described by management thinker Peter Senge in his iconic book, *The Fifth Discipline* (published in 1992). In this book, Senge describes how any outcome results from the inputs and processes that drive the outcome — just like our pro golfer example above.

Systems Thinking can be described thus:

> "The outcome that is achieved from any
> process is a direct result of the policies,
> procedures, measures and reporting that
> manage that process"

In the case of inventory, think of the outcome as the problems that are created with inventory, such as the existence of excess inventory or a shortage of inventory. These problems are the direct result of the policies, procedures, measures, and reporting that are applied to managing that inventory — these are the inputs and processes that drive the inventory outcome. Taking a Systems Thinking view guides you to a more complete review of the things that you must influence if you are to improve your inventory results. For example, many people use optimization software to recalculate their ROP and ROQ. However, as explained previously, this approach ignores the systems effect of other inputs and processes that influence our results.

Figure 4-1 is a simplified version of the inventory supply chain showing the input of supply and the pull of demand on an SKU. In practice it is easy to ignore the factors that influence the supply, demand, and SKU management, but this would be a mistake. Factors of supply, such as the lead time, frequency of ordering, quantity ordered, the timing of delivery, and price are not fixed. Similarly the internal processes for SKU management and factors of demand are not fixed. It's just that sometimes it appears to be more convenient to assume that they are. This is thinking that ignores the systems in which inventory management must operate.

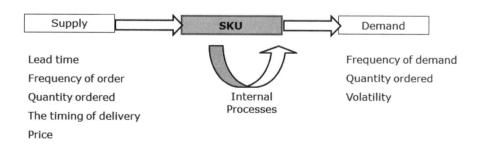

Figure 4-1 Simplified Supply Chain

The MIM Cycle in Detail

L et's expand on the process discussion by turning our attention back to the MIM Cycle diagram introduced in Chapter 2. This is replicated here as Figure 4-2. For each step of the process there is a feedback loop that represents the actions that are taken at that point in the process. This feedback not only informs that step, but also influences the results of subsequent steps and the results that are ultimately achieved — this is the essence of Systems Thinking. To fully understand the influence of the steps in the process, let's break this diagram down into its basic elements.

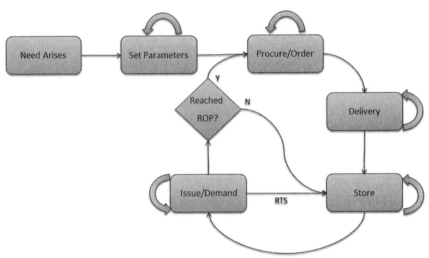

Figure 4-2: MIM Cycle

Need Arises

There will always be some initiating event that results in a need to consider holding inventory. For engineering materials and spares, this might typically be the installation of new equipment. It might also be the recognition that a particular spare is required, if it were not previously recognized. One influence here is the choice of equipment that might be different to that already in place, resulting in a need for different spares. Another influence is the timing of the recognition of need. Is there time to plan or is the need only recognized when a breakdown occurs? This will obviously influence the extent of review for the next few steps. In an emergency, we place more attention on getting what we need and less attention on whether we are buying the right quantity for future needs.

Planning and Set Parameters

As a result of the initiating event, there will need to be some planning and the setting of parameters for managing and ordering the parts. At this point in the process, estimates are made of the quantity that might be used and the frequency of demand. For engineering materials and spare parts, inputs should come from designers, engineers, and maintenance planners. The results ultimately achieved will be very dependent upon whether these groups have input and/or responsibility for the outcome. With engineering spares we often rely on the data provided by the equipment suppliers, but their data is usually generic and does not take into account your situation. Unfortunately, spares management of new equipment is usually a last minute thought that is poorly defined and managed. Without good planning and appropriately set parameters, the result will be the establishment of parameters that commit the organization to over- or under-stocking of parts from day one.

Procure/Order

Once the parameters are set, an initial order is placed. In most systems this is also the first of the recurrent elements — that is, further orders are placed as required. As this is the first recurrent element, it is not typical that the reorder parameters are revisited, yet needs may have changed. Also, there may be procurement procedures that dictate the way in which the items are purchased or the quantity purchased. The impact of these is not always reviewed in the context of the total system — just the individual procurement 'silo' in which the procedures are written. Containing the approach to the individual 'silo' can result in ongoing overstocking. Another issue with this step is timing. If the time it takes to internally process an order is excessive, then the quantity of parts required to be held in inventory will be greater because the process time increases the overall lead time.

Delivery
Just as ordering timeliness is an issue so is the timeliness of entering a delivery into the inventory management system — this is in addition to the time it takes for the supplier to deliver the order. Consider what happens when an order is delivered to your site. Is it quickly processed and entered into the system so that it appears on the computer as being available? Or does it wait some time until this important administration task is completed? You should not assume that items are quickly processed.

Store
How well is the store room set up? Is it easy to find items? Is the inventory count accurate? Are items put away in correct locations? Are items stored in a manner that preserves them so that they are fit for purpose when required? Are item removals accurately recorded? Each of these issues has a big impact on your inventory management results.

Issue/Demand
What behaviors are evident in the demand for parts? Chapter 7 discusses the phenomenon of induced volatility, which occurs when people take more of an SKU than they really need at that time and, as a result, the data used for calculation of holding parameters are misleading. When the additional items are later returned to store, it also has an impact on the data (see *Return To Store* below).

Reached ROP?
How quickly are the need to reorder recognized and the purchase requisition raised? A slow reaction will result in greater inventory holdings.

Return To Store
As we saw from Figure 2-3, improper management of the Return to Store process can have a significant impact on the quantity of an SKU that is held. The impact comes about from two different effects. First, if the items have already been reordered, then the stock that is subsequently held will be greater than needed or planned. Second, the removal of items without corresponding need (hence the subsequent return to store) means that the data that are used for calculation of

holding parameters will be misleading. Use of this data will result in systematic overstocking.

Once Systems Thinking is understood, the influence of process on materials inventory outcomes is easy to see. Addressing process issues must then surely be the most likely approach to drive lasting and economically sustainable materials and spares inventory management outcomes.

Chapter 5

Policies & Procedures

If people are the key players in materials and spare parts management, then the policies and procedures are the rules of the game. It is the policies and procedures that communicate the expectations and set the guidelines for action. Most sophisticated organizations will have corporate policies (those high-level guidelines that operate as a statement of intent) and operating level procedures (how to reorder items, conduct stock takes, etc.); both are important. However, the missing element is usually the specific guidelines on how much to order or hold (or at least how to decide that value) and how to store items to ensure their integrity. There is some irony in this because these missing guidelines are the ones that influence the actual level of inventory. Whether your goal is availability or minimal investment, the level of inventory is your most important factor.

This section discusses the development of a spares stocking policy, the need for categorization of spares, and that most important of all questions, 'when is critical really critical?'

When Is Critical Really Critical?

There is an old saying that goes something like 'a maintenance engineer never met a spare he didn't like.' But perhaps maintenance engineers should not like their meta-spares quite so much. What's a meta-spare? I'll get to that.

In my work helping companies optimize their engineering materials and spares inventory management, I am almost always greeted at a new client with the exclamation 'you can't do anything about that spare; it is critical.' Critical spares are held in high regard and treated as untouchables. However, whether or not a spare is critical is not the point when it comes to inventory review. You can still hold too much of a critical item. In my opinion, the comments I hear on critical spares are usually more emotional than scientific. Let me explain that.

The goal of any maintenance program is to minimize the frequency and consequence of equipment failure. Spare parts availability plays an important role in minimizing the consequences of failure by reducing the downtime associated with the failure. Spares, of course, do nothing to impact the actual occurrence of the failure. (Yes, we could have a discussion on the quality of spares used so let's just assume that here we are discussing parts of suitable quality.)

A critical spare can be defined as a component that, if unavailable, would prevent the plant from operating. So if the absence of that item would prevent the plant from operating, then the item is critical. In most plants, however, this definition covers a wider number of items than those defined as critical within the plant's inventory system.

The first point to recognize is that truly critical spares are likely to be far more widespread than your system currently defines.

Engineers will happily (well, not happily) review items that are not classified as critical but which actually are critical, just not classified as such. Yet they will shy away from reviewing items that are officially defined as critical. In this case, any review is driven by the classification rather than the opportunity or cost benefit. Nobody would use this approach in reviewing any other capital investment.

The second point is that not all items classified as critical are genuinely critical. In a recent example, a company had 24 compressor valves defined as critical items. It was true that the compressor would not run without these valves. However, the company separately held pins, plates, and springs for these valves. In practice, this meant that the valves were never needed. If there was a failure between planned valve maintenance activities, then the pins, plates, and springs would be used, not the spare valves. Based on this practice, it could be argued that the pins, plates, and springs were critical, even though they were actually not classified as such.

Perhaps what this really means is that the definition of critical needs to be expanded to include the words 'for which there is no viable alternative.'

Therefore, our definition of critical is now 'a component that, if unavailable, would prevent the plant from operating and for which there is no viable alternative.'

No matter what definition is used, classifying an item as critical tells us that we need to have ready access to the item when it is required. It doesn't tell us how many need to be carried, nor does it tell us whether they need to be owned prior to needing the item.

Let's examine those statements for a moment.

Ready access: means that it is available when we need it. It doesn't mean that it must be in our store or on our site but that we must be able to readily access the part when needed, in a timeframe that is acceptable.

It doesn't tell us how many we need: classifying the items as critical tells us is that we will need the item. Whether we stock more than the minimum required is dependent upon the usual usage, ordering, and supply issues.

Nor does it tell us whether they need to be owned by us: classifying an item as critical says nothing about ownership. Why can't someone else own the item just as long as we control the item?

What about the second spare?

The management of engineering materials and spare parts (in particular) is typically conservative, so let's say that you really do want to have the item in stock in your store room, sitting there waiting for you. Let's look at some further issues that you can address.

The decision to hold that first spare is usually based on the determination that the spare is critical and, therefore, required. However, what about the second spare? The decision to hold a second spare is based on the probability (and consequence) of a failure during the time it takes to restock the first spare. This is of course a very limited time. The requirement for a third spare is based on the probability of two failures during the restocking period of the first spare. (The Poisson Distribution in Chapter 2 shows the probabilities quite well.)

Let's call these second, third (and subsequent) spares, meta-spares — literally spares for your spare. The second spare that you hold is a spare for your spare. The third spare is a spare for your spare's spare. As you can see the function of these spares is to be available while your first spare is restocked.

So, if critical really means 'available', there may be a number of actions that you can take that allow you to have availability but with lower costs and fewer meta-spares. These actions might include consignment stocking, identifying effective duplication, and reviewing supply chain opportunities.

In another example, a company that had steadfastly used sea freight for overseas-sourced components (because the freight cost is lower) realized that the total cost of ownership for some items was less if they used airfreight. This was because they could then hold fewer meta-spares. They still had ready access to the items; they just changed their supply chain.

When most engineers say to me 'you can't do anything about that spare, it is critical' what they are really suggesting is that because the spare is classified as critical they cannot hold any fewer than they already do. Yet as we have just seen, this cannot be allowed to go unchallenged. If the above arguments don't convince you to include a review of your critical spares in any spares review, consider the following.

Engineers often justify critical spares by saying that the cost of failure so outweighs the cost of holding the spare that the cost of the spare doesn't matter. Although based on a degree of financial sense, this is largely an emotional argument. Logically, when you reach 100% service coverage you don't need any further spares. Unfortunately many companies hold more than they need for 100% service coverage.

Figure 5-1 represents the confidence level of spare parts availability plotted against the number of spares held. In this example, if the company holds X number of spares their confidence of availability will be 100%. Obviously there cannot be greater than 100% so if they hold more than X spares the confidence level will still be 100%. Therefore, purchasing more than X spares is a total waste of money! It doesn't even matter whether or not the spare is critical; holding more than the 100% level produces no benefit.

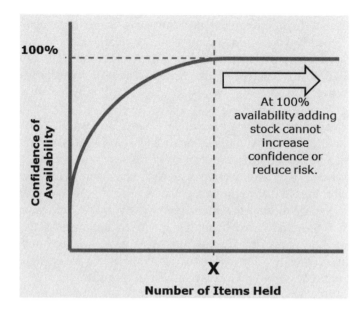

Figure 5-1: Confidence Level for Spares Availability

Cost Benefits are Not Universal

The cost benefit of holding spares is not universal, even for an individual item. The probability of usage reduces for each subsequent spare. The additional benefit diminishes to zero as the number of spares increase.

Still not convinced that you should review your critical spares?
Consider these issues.

- What if the spare is no longer critical? Perhaps the operational needs have changed but no one has updated your data. Should you still carry just as many, or any at all?

- What if the spare is no longer usable? Perhaps a review will identify if the spare itself has deteriorated through a lack of appropriate care. Imagine needing the critical spare and finding that it won't perform!

• What if the equipment that the spare is for has been modified? Recently I saw a situation where a gearbox could no longer be used in its intended location because the connections and motor had been changed. This gearbox was still regarded as a critical spare!

So it appears that not all critical spares are equal. Not all spares classified as critical are actually critical, and some spares not classified as critical may actually be critical!

Confused? For engineering materials and spare parts management, this confusion means that defining a spare as critical is not an excuse for not including the spare in your inventory review. Even with critical spares, it is possible that too many are being held. With appropriate thinking and judgment, critical spares may actually provide a significant opportunity for your inventory review. Furthermore, a review of the associated inventory management processes could deliver significant benefits.

The Spares Stocking Policy

Companies these days are awash with policies. They have environmental policies, safety policies, purchasing policies, contract management policies, audit policies, corporate policies, expense policies, travel policies, maintenance policies, and some even have internet usage and blogging policies. When it comes to engineering materials and spare parts management, companies often have inventory policies, stock taking policies, cataloging policies, warehouse policies, issues and usage policies, and disposal policies.

With all this corporate red tape, why would you need (or want) one more policy? The answer to this question lies in the benefits you get in return.

Imagine an organization that spends millions of dollars on materials to support their maintenance activities. Imagine having no budgetary limits or constraints on how much they spend. Imagine a situation where the only time this is an issue is when someone finally says 'Hey, what's going on here?' Imagine that when that someone finally notices, it is too late — the money has been spent.

Sound familiar? It should because this situation describes the engineering materials and spare parts management of many industrial organizations today. (Of course they have budgets for actual maintenance expenditure, but not often for how much inventory they carry. Recall also the discussion in Chapter 3: The Financial Impact of inventory.) The one policy that helps to sensibly control and contain this expenditure is a Spares Stocking Policy. The benefits for having a well thought out and well-defined Spares Stocking Policy (that is implemented) include:

- Preventing the excessive purchase of items
- Improving the timeliness of availability of items
- Reducing the level of obsolesce and write downs
- Faster recognition and clean up of the few items that do become obsolete

In short, a Spares Stocking Policy saves money through reduced expenditure and improves reliability through reduced downtime.

Now that we can see how important this policy is, let's look at how to develop a high performing parts policy.

A policy can be defined as 'a deliberate plan of action to guide decisions and achieve rational outcomes.' So a Spares Stocking Policy will provide your team with guidance on how much and when to purchase in order that you achieve the availability you need without over-expenditure. In addition, by comparing actual outcomes to the policy, you can use the policy as a key part of the feedback to identify problems and issues with engineering materials and spare parts inventory management. Your policy provides an internal benchmark against which you can compare performance.

Your Spares Stocking Policy will not tell you what to purchase. That is, after all, an engineering decision based on your equipment needs and selected approach to maintenance — your maintenance policy! What your Spares Stocking Policy must do is:

1. Identify the categories (or types) of engineering materials and spare parts inventory that you carry
2. Provide guidance on the quantity to carry and how to calculate it (based on the above categories)
3. Provide guidance on the timing of purchases (for example, with shutdown programs or routine replacement)

4. Identify storage issues and/or locations
5. Provide purchasing and/or price information

Let's look at these individually.

1. Identify the Categories

It is important to realize that not all Spares Stocking Policies are the same. The policy must be specific to the spare category and your situation. The key to developing your categories is that they must be MECE — that is, Mutually Exclusive and Completely Exhaustive (pronounced mee see). Let me explain that.

Mutually Exclusive means that each category must be independent to the others in a way that items cannot be assigned to more than one category. For example, if you used a category called Breakdown Spares and another called Fast Moving, these two categories are <u>not</u> mutually exclusive. A spare part could be required for breakdown repairs and, if used regularly, could also be fast moving. This confuses your use of categories and limits your opportunity to review and analyze your spares holding. You can see from this example that the categories must be linked conceptually. A breakdown category refers to an event whereas a fast-moving category refers to frequency of usage. Mixing the concept in the definition results in poor categorization and, therefore, limits the benefits of categorization.

Completely Exhaustive refers to covering all of the possibilities. Continuing the above example, if you only used Breakdown and Fast Moving, aside from not being mutually exclusive, these two categories have clearly not covered the slow-moving items, so there may be items that do not fit either category. To be Completely Exhaustive, your categories must cover all possibilities.

Some types of categories that you might consider include:

- Spare Parts: Parts that, if unavailable, would not prevent the plant from operating and/or where usage could be planned. This might include subcategories such as fast-moving parts that turn over regularly, slow-moving parts that turn over occasionally, routine maintenance parts that are used only for specific regular maintenance events (and therefore can be planned), and project spares that are placed in inventory in preparation for specific projects which are one off events (and so can also be planned).

- Critical Spares: These have previously been defined as components that, if unavailable, would prevent the plant from operating and for which there is no viable alternative. These spares may also be subdivided as fast, slow, or routine.

- Consumables: Likely to be very high usage, but relatively low-cost items that may be directly expensed to a cost center as a means of minimizing the administration associated with the level of usage. The key attribute of a consumable is that it is consumed when used and, therefore, can only be used once.

- Bulk Commodities: Items that are essentially consumable, but where their supply, storage, and usage requirements set them apart from other consumables (such as fuel and oil).

- Capital/Insurance Spares: Capital/Insurance spares are usually very high value items that are peculiar to a specific asset. Because of this, they can be defined in the accounting system as Property, Plant, and Equipment rather than inventory (ask your CFO about this definition). The term Capital Spares is also often used to cover spares supplied as part of an initial capital purchase rather than those purchased subsequent to the capital purchase. In either case, they usually are not counted as part of the financial value of inventory, but they do need definition in order to clarify what they are and whether they should be counted.

You can see that these categories are MECE because they are initially divided on the basis of unique attributes. It is, therefore, unlikely that any one item could be included in more than one category. Also these categories cover all potential inventory types. Figure 5-2 shows a simple decision process to help identify into which of these categories an item may fit.

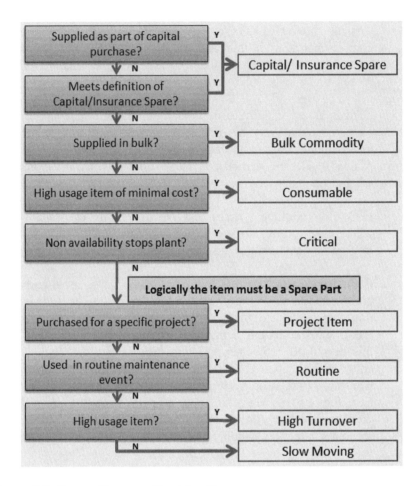

Figure 5-2: Spares Category Decision Tree

Now that the categories are defined, you need to think about the other aspects that make up the policy.

2. Provide Guidance on the Quantity to Carry and How to Calculate It

The quantity to carry is the issue that gets the most focus in inventory reviews; a number of different techniques can be used to calculate this quantity. For example, you may choose to use software to calculate your requirements and, therefore, base your policy on the output of that software. As has been discussed extensively in other chapters, this approach has both pros and cons. On the one hand, the software is convenient. However, if your data is inaccurate or incomplete, or you do not fully understand the assumptions in the algorithm used, and the constraints that it imposes, then you are likely to overinvest, create a poor inventory mix, and sub-optimize your outcomes. Alternatively, you may choose to use a process-based approach (such as Inventory Process Optimization™ — see Chapter 9); this approach overcomes the data issues, assumptions, and constraints by tackling them head on and providing guidance based on optimizing your management processes. Whichever approach you choose, your Spares Stocking Policy must clearly state the process or rules to be used.

Here are some further rules of thumb to look out for and comments that might help in determining the quantity to carry and how to calculate it.

- The less critical the part and the shorter the lead time, the fewer items that need to be held.
- Conversely, the more critical the item and the longer the lead time, the more items that need to be held.
- You should always strive to hold the minimum quantity required, don't manage to the maximum.
- The more predictable the usage the better the delivery can be timed to suit usage.
- Be aware that data does not always reflect what you think it does.
- Demand for parts is often driven by the behavior of the team rather than the failure characteristics of the equipment.
- Similarly, supply lead times are often driven by the procurement, receiving, and storage procedures.
- Sometimes 'calculation averaging' leaves you short of the one-time usage requirements.
- You do not necessarily have to fill all the storage space (for example, with bulk commodities).
- Simple protocols and criteria make both the initial decision and the subsequent review of that decision easier.
- There is no 'one size fits all' calculation that can be applied. Everyone's circumstance and process is different, and so requirements are different.

3. Provide guidance on the timing of purchases

For some categories it is sufficient to set your quantities via an ROP and ROQ. However, for the categories that relate to parts where there is a degree of predictability in usage, it is important to provide guidance on the timing of purchases. For routine maintenance or project spares, the planned timing will be known in advance. Therefore, your policy should include setting the criteria for ordering and delivery. For example, you may say that all shutdown spares should be ordered for delivery to occur in the month preceding the shutdown. This criterion can be applied no matter whether the part has a six-month lead time or a one-week lead time. The important principle here is that the cost of inventory includes the time value of money for the time that you hold it. The shorter the holding time, the lower the overall holding cost.

4. Identify storage issues and/or locations

It may be that your storeroom is organized in a way that separates the inventory categories. It may be that some of your spares require special treatment for their own care and to improve their longevity. In either case, it is worthwhile setting out your expectations so that people will not have to guess!

5. Provide purchasing and/or price information

Similar to number 4, your policy may include guidance on procurement or pricing. For example, some items may need to go through a central contract arrangement whereas others may be purchased locally. Making these requirements clear will help ensure that the right price is paid for the items that are purchased.

A Checklist for Development

Figure 5-3 shows a checklist that you can use to review your progress in developing your Spares Stocking Policy. This is a quick and easy tool to see whether you have all the issues covered.

In addition, here are a few other thoughts on setting your Spares Stocking Policy.

- Your Spares Stocking Policy does not have to be a highly detailed and complex document. As a starting point, try setting a few simple guidelines.

- The best time to determine your policy is when you are establishing your storeroom or inventory. If you already have these in place, then the best time to determine your policy is now!
- The responsibility for establishing a Spares Stocking Policy lies with the Inventory Manager or whoever controls the inventory management processes. However, it is recommended that you involve all of the key people that influence your inventory. This includes: storeroom personnel, maintenance, engineering, procurement, finance, and inventory management.
- Any policy is worthless if it is not followed. Therefore, you must be prepared to communicate the new expectations and follow up on its use.
- Once your policy is developed, compare your current holdings to your policy in order to identify gaps and opportunities.

Spares Stocking Policy Checklist

❑ **Categories defined**

❑ **Categories are MECE**

❑ **Simple and defined process for calculating stock holding based on category type**

❑ **Guidance on timing of purchases**

❑ **Identified storage issues**

❑ **Purchasing/pricing information**

❑ **Involvement of all key personnel**

❑ **Policy communicated**

❑ **Compare current holdings to new policy**

Figure 5-3: Spares Stocking Policy Checklist

You Need To Be More than Just a Good Administrator

The key reason to develop a Spares Stocking Policy is that it sets the expectation of how and when inventory is to be ordered and how it is to be stored. The Spares Stocking Policy is different from your inventory management procedures as these procedures usually relate to the administration of the inventory — that is, who does the ordering, how items are booked out, and so on. Many organizations have good and robust inventory management procedures and, as such, are good inventory administrators. This is not, however, enough. It is only by the adoption of an appropriate Spares Stocking Policy that companies become good inventory managers.

Previously in this section it was stated that a Spares Stocking Policy saves money through reduced expenditure and improves reliability through reduced downtime. This alone makes a Spares Stocking Policy the one policy that you may not have but that you really do need.

Chapter 6

Issues, Myths, and a Few Home Truths

Now that we have worked through a recap of the basics and understand the impact of people, process, policy, and procedures, we need to consider some of the impediments to inventory management success. These are a combination of issues, myths, and home truths that seem to be forgotten. An understanding of these issues, myths, and home truths is of enormous value in working your way through potential solutions and their application to your situation. This understanding enables you to address objections by your team, evaluate solutions put forward by vendors, and recognize behavioral issues that need to be addressed.

Why Inventory Reduction Often Fails

Before reviewing what works in materials and spares inventory management, let's first understand why efforts in the area of inventory reduction often fail. There are two key reasons that the usual approaches to inventory reduction fail to deliver sustainable results:

1. The focus is often on high-priced items that are seen as providing the greatest impact.

2. The focus is on high turnover items as any change in this area is likely to be realized more quickly.

Both of these approaches are flawed.

The first approach neglects the total investment made in a line of inventory, that is, the total dollar value invested — the value per item multiplied by the number of items. The total dollar value invested is important because it represents how much cash is tied up, not how much the item costs. If the goal is to reduce the inventory investment, then you must always follow the cash.

The second approach — focus on high turnover items — ignores slower moving items and often results in an imbalance of inventory. How often have you been faced with having plenty of inventory, but none of the right items in stock? Correcting a focus on high turnover items requires a process aimed at the cash investment rather than just the inventory activity.

Cracking the Inventory Code

Materials and spares inventory management brings with it a whole new language and a dizzying array of acronyms – ABC, EOQ, ICR®, IPO, GSE, JIT, MRO, MRP, SOS, SKU, RCS, ROP, and ROQ to name just a few! And like all jargon and acronyms, the meaning and purpose is not always clear. So, how do you know what is what? How do you determine if a suggested solution is right for your situation? How do you bridge the gap created by jargon to ensure that you resolve your materials and spares inventory management issues in a way that is sustainable? To do this, you need to crack the inventory code. Cracking the inventory code is not about focusing on acronyms, nor does it require a secret code book. It does require an understanding of a few inventory home truths.

Home Truth #1: Inventory Does Not Exist in Isolation

Often inventory is treated as if it exists independently of other operational and environmental factors. However, this is not the case. For example, the level of engineering spares inventory that is actually carried (as opposed to the quantity recommended by the stocking policy) is influenced by a wide range of factors, most of which you can change and some that you cannot. Some of these factors are:

- The equipment types and number in service (difficult to change)
- The plant network (difficult to change)
- The equipment reliability (improving this is the aim of your reliability program)
- Inventory and other company policies (can be changed)
- Inventory, purchasing and maintenance procedures (can be changed)
- Human behaviors (can be changed)
- The supply chain (can be changed)
- Supply chain reliability (can be changed)
- Reporting structures (can be changed)
- Storeroom management (can be changed)
- Metrics and measurement (can be changed)

Someone once said that 'all problems end up in inventory.' What they meant by this was that problems with any of the above factors can result in excess inventory holdings. The key is to realize that by changing some of these operational and environmental factors you will have a significant, positive, and lasting impact on your inventory stock levels. Changing only the planned levels does not drive these outcomes.

Home Truth #2: There is No Universal Solution

The next thing to understand is that indirect inventory has characteristics that make it different from the direct inventory that is held for making products for sale (e.g., raw materials, WIP, and finished goods). A major trap for most people seeking to improve their engineering materials and spares inventory performance is that they fall back on solutions that are applicable to the relatively stable and predictable demands of direct inventory.

For example, let's look at the philosophy of Just In Time (JIT), perhaps the most widely known inventory management technique. The JIT approach was developed in a production environment to improve process flow and eliminate the planning function. JIT relies on the use of a Kan Ban to regulate demand and supply so that products are 'pulled' through the manufacturing process based on demand at the next stage. This focus on process full and demand pull makes the JIT philosophy an all-pervading aspect of production management because it influences the way people think and behave, and the way that they manage their processes. Contrary to popular belief, JIT is not a technique but an outcome that results from the application of other techniques. It is not a universal solution for inventory reduction.

In addition, it is important to remember that inventory is not a single homogenous 'thing.' Inventory is made up of many hundreds or thousands of largely independent SKUs. Treating inventory as a singular item is a bit like treating other assets as a single item. For example, nobody has a plan to maintain a power station, they have plans to maintain the components that when assembled together become a power station. Each component is different and has different characteristics and needs. It is exactly the same with inventory. It makes no sense to apply a single tool or technique to the entire inventory as if it were a 'thing' and to think that it will resolve all issues or produce the best overall result.

To crack the inventory code you need to ensure that the tool or technique being considered is applicable to the type of inventory you are reviewing and is not just a so-called universal technique.

Home Truth #3: There Are Only Two Outcomes of Concern

Although inventory is influenced by many factors and is made up of many different components, it is important to realize that at a fundamental level there are just two outcomes that are of concern. There is either too little, resulting in stock outs; or there is too much, resulting in overstocking, slow moving stock, and obsolescence. Often people will say that they have a mix problem, but all this really means is that some items are overstocked and some items are understocked.

Overstocking and understocking are obviously two very different outcomes, but it may not be so obvious that the solutions for each may also be different. For example, overstocking could be a result of setting incorrect inventory levels. But is also almost certainly a result of the influence of the operational and environmental factors mentioned previously. Understocking could be addressed by reviewing the above factors, but will almost certainly be improved by recalculating the required ROP. The emphasis is quite different. You cannot apply one approach and expect it to resolve both problems. Yet this is precisely what people do!

To crack the inventory code you need to identify the outcome you are trying to influence and apply the appropriate solution. In engineering, no one would advocate driving in a screw with a hammer, but this is the equivalent of solving an overstock problem with a technique better suited to an under-stock problem.

Home Truth #4: The Important Characteristics Are Not Always the Most Obvious

When managing inventory it is convenient to segment the inventory using descriptions such as fast- or slow-moving as this helps with setting stocking policies (see Chapter 5). However, these segmentations often focus on the frequency of demand. Although frequency is an obvious characteristic, it is not always the most important characteristic. When reviewing minimum stockholdings, a more significant factor is the variation in the quantity of each demand. This less obvious factor tells you which review approach to use.

For example, if an item, such as an electric motor, is only ever used as a single item, then the appropriate holding should be calculated using a Poisson function. If the item has demand that varies (say 1 this time, 5 next time, 3 the time after), a Gaussian function is more appropriate. It is not the obvious characteristic of how frequently the item moves that counts, but how many might move at each demand. (Both of these functions are discussed in Chapter 2).

Every Company is Unique

Every company's inventory situation is unique because of the combination of issues that impact their inventory. For example, a steel producer in Illinois and a steel producer in Southern Australia face unique challenges because of their location, proximity to suppliers, local culture, age of plant and equipment, workforce relations, physical layouts, management style, and so on. Equally, a parts supplier is different from a parts manufacturer because of where they sit in the supply chain. A company that is overstocked faces different challenges to one that is understocked. In all of these examples, the characteristics that you must focus on vary, depending upon your purpose and intent.

It is by understanding these home truths that you can avoid many of the issues that result in underperforming materials and spares inventory management. Better than that, though, by cracking the inventory code, you can identify which of the range of possible solutions are best suited to your situation.

5 Myths of Inventory Reduction

In Chapter 3 it was established that a program of inventory reduction can be one of the most powerful and value-adding activities that a company can undertake. This is because inventory reduction generates cash, just as sales or cost reduction activities generate cash. This cash is just as real and just as valuable to the company as cash that is generated through sales or cost reduction.

When the inventory being held is indirect inventory (that is, it is not being held for manufacture and, therefore, automatically moving through the supply chain), then the benefit is even greater. With engineering materials and spare parts inventory, it is possible that some inventory will never be used and will only ever be a cash drain on the company. So why do so many companies allow their engineering materials and spare parts inventory to be a 'fat and lazy' investment? Why do they not apply a simple process that safely minimizes their investment? There may be several reasons.

First, there may be limited knowledge of the alternatives. Many companies think that optimization using software is the only solution. However, companies that do apply optimization software may be achieving less than 1/7th of their inventory reduction potential.

There may also be a lack of resources to conduct an inventory reduction program. However, in the author's experience, once the potential to generate cash with zero capital investment is understood, then the resources can always be found.

However, it is most likely that companies are prevented from taking action because of the beliefs and assumptions that they make about inventory. These beliefs and assumptions are truisms that can (and do) destroy a company's wealth by allowing an over-investment in inventory. Five common truisms have been identified; these are called the 5 Myths of Inventory Reduction. To successfully effect an inventory reduction program, a company must recognize these myths and challenge them whenever they are raised as the reason for inaction or lack of progress.

Like all good truisms, they are each based on an element of truth, but they are not universally true. And like all management myths, they work to prevent effective action. The impact of these myths is that they limit the ability to fully realize the potential opportunity of inventory reduction. Therefore, they limit the cash that may be realized through delivering a successful program. Recognizing these myths and applying appropriate management solutions to overcome them will help you to achieve true inventory optimization.

The Five Myths of Inventory Reduction are:

1. Economic quantities save money.
2. Risk must be re-evaluated to reduce inventory.
3. Consignment stocks must cost more.
4. Software will solve the problem.
5. Putting items into inventory shares the cost.

Myth #1: Economic Quantities Save Money.

In engineering materials and spare parts inventory management, items often get ordered in an economic quantity so that the cost per item is at a minimum. This way of ordering is seen to be economic because the subsequent issue cost of the item is at a minimum and the business, operational, or project budget subsequently records a lower cost. The term *economic order quantity* is often used.

Ordering in this way is not economic, however, in situations where the items are not used, where they are written down as slow moving, or where the holding cost ultimately exceeds the procurement saving.

Determining the true economic position of holding spares requires a consideration of the total company cash cost, not just the departmental or project charge.

Example

In a manufacturing operation, it is decided that a special widget is needed as a spare. The set-up costs for making the widget are such that the first widget would cost $2,000. However, once set up, the supplier will provide five widgets for $3,000.

If five widgets are purchased, the purchase cost would be $600 each — an apparent saving of $1,400 over the single widget cost. On the surface, this prospect is quite appealing.

But what if the other widgets are not used and are written down as slow moving after, say, 4 years? The so-called economic approach really costs the company the original $3,000 plus the annual cost of holding the four widgets in inventory (at a conservative 20% per year this is $2,400 x 20% x 4 years = $1,920). The total cost over the four years could be as high as $4,920.

Therefore, while the operational budget showed only a $600 expense when the only part that was needed was issued by the storeroom, the company incurred a $4,920 cost. Purchasing just a single widget would have cost the company just $2,000.

Furthermore, in the event that the $2000 widget was seen to be too expensive, an alternative solution might have been found.

Myth #2: Risk Must Be Re-Evaluated To Reduce Inventory.

Reducing holding quantities in inventory is often seen as requiring a corresponding increase in operational or financial risk. The risk might be the risk of a lost sale or, in manufacturing, the risk of extended downtime. How often have you heard someone say 'We need our inventory or we won't make sales' or 'We need our inventory or our downtime will go through the roof'? Some companies believe that inventory can only be reduced when their maintenance systems are sufficiently sophisticated that they can predict demand or they have eliminated unplanned failure. (Many consultants and vendors also work hard to perpetuate this myth.)

Both of these positions *implicitly* assume that the existing holdings are as lean as they can be in the current operational dynamic. Although it is possible that this is true, experience shows that it is unlikely. Consider some of the issues that might result in a need to change inventory holdings:

- The initial parameters were set without any history of usage and, therefore, may not have been correct. Typically, incorrect settings are only reviewed when a stockout occurs. A stockout is when there is demand for an item, but none in stock. An overstocked item may not ever have been reviewed.

- Suppliers have implemented supply chain initiatives. In the past ten or so years, supply chain improvement has been the focus of most distribution activity. Initial assumptions about supplier capability may now be outdated and may not have been revisited, especially if the capability of your inventory management software to recalculate values is turned off.

- Continuous improvement in your own systems. Initial assumptions about your own capability and demand requirements may have changed in an incremental way through continuous improvement. For example, improvements in your own receipting process may not yet have translated into lower holdings.

- Opportunities for consignment stock. Shifting market dynamics may have resulted in new opportunities for consignment stock that were not previously available.

These are just some of the issues. In each of these examples, the actual risk will have reduced significantly compared to the initial assumptions. If the tolerance for risk has not changed, then reviewing these issues will, at worst, reset the system to the tolerable level of risk. In the case of consignment stock, the risk level should remain unchanged. (Note that there is a difference between the level of risk and the level of direct control. Consignment stock reduces direct control.) Addressing these issues requires a consideration of the demand and supply dynamics and how these dynamics may have changed over time.

Myth #3: Consignment Stock Must Cost More.

Arranging to pay for items only at the time they are issued for use is referred to as consignment stocking. With consignment stocking, the supplier owns the items, even on your premises, until your team issues or uses them. Typically a monthly review of the quantity issued drives invoicing.

As the supplier must now finance the stock and accept the inventory risk, it is often believed that additional costs will be passed on to the purchaser. This is not, however, always the case. Gaining control of stocking gives the supplier many more options to be proactive in the management of the supply chain. For example, they can schedule manufacturing and deliveries to suit their timetable rather than be reactive to your purchase orders, or they can draw on a wider network to manage safety stocks. They may even decide to draw against your holding to supply other customers. The flexibility of consignment stocking can provide the supplier the opportunity to reduce supply costs through improved manufacturing and supply chain efficiencies.

Spares suppliers make their profit from the sale of material. They don't want stock sitting idle. Therefore, they will have a greater interest in redirecting items that become slow moving. As a supplier, they obviously have direct access to the market in order to arrange the sale of items that are not moving at your site. This greater market access enables a degree of flexibility in managing stock that is not readily available to the typical end user and reduces the carrying risk of the inventory.

Some suppliers will try to charge more for consignment stock, citing the investment cost and risk as the key reason. However, a sensible approach to consignment stocking that identifies the supply chain opportunities can result in the double benefit of reduced inventory and reduced costs. (See Chapter 8: The 7 Actions for Inventory Reduction).

Example

In this example the user company held a large supply of hydraulic hoses. They had tried using a call-out service to provide a hose when needed. However, they found this arrangement unsatisfactory because they sometimes had to wait too long for the service van to arrive.

In negotiating a consignment deal, the supplier agreed to consignment stocking on the basis that:

- The company would monitor holdings (and would be responsible for shortages).
- The supplier could restock at its convenience (hence, it became fill-in work).

- They could draw on the stock for other local emergencies if required.

Under these arrangements the supplier was able to gain several efficiencies in working with the company. As a result, it was not under pressure to increase its prices. In this win-win situation, the user company gained a $100,000 inventory reduction with no increase in their risk.

Myth #4: Software Will Solve The Problem.

Almost everybody realizes that software alone does not provide a solution. Yet, many companies, when faced with an inventory reduction program, see the need for new software as a key prerequisite. Sometimes this software is inventory management software and sometimes it is the use of a new tool such as optimization software.

Software provides data availability and visibility. Both are key requirements of an inventory review program, but the software itself is only a tool. Like all tools, it needs to be used properly and in the proper context. Ongoing inventory reduction is achieved by a combination of culture, knowledge, and data availability.

Here is the problem: if addressing the culture and approach to materials and spares inventory management can drive an improvement in performance, then conceding that these need to be addressed is the equivalent to admitting to an initial failure on the part of management. Psychologically, this type of admission is difficult and can be both a challenge and a humbling experience. As a result, a new tool is seen as the way to effect change as this provides the reason that new outcomes can be achieved (that is, because we have new tools, not because we didn't originally do our jobs properly).

There are a number of examples where the same software exists in different parts of the same company, under the same policies and procedures, and yet vastly different results are achieved. There are also examples where companies buy a new optimization tool, but achieve no sustainable change. Clearly the issue is not the availability of the software, but rather the culture and measures within which it is applied.

By understanding the dynamics of materials and spares inventory management, and providing appropriate measuring and reporting systems, a company can ensure an ongoing focus on inventory reduction. This ongoing focus will last long after the software implementation is forgotten.

Myth #5: Putting Items into Inventory Saves Money.

Adding an item to inventory is sometimes seen as a way of spreading the cost of the item so that the original purchasers can get a lower charge to their budget. Managing budgets in this way is particularly relevant with project and engineering items that have a minimum order quantity.

Myth #5 is similar to Myth#1: Economic Quantities Save Money, except that the focus here is not on purchasing efficiencies, but rather on operational or project budgets. Ordering items where the delivery is in excess of needs and having the excess put into inventory reduces the direct cost to the immediate budget. Managing the purchase in this way has the impact of appearing to save money, but it does not change the actual cash cost to the company.

A mismatch between the authority to assign items to inventory and the responsibility for the inventory investment can result in inventory being used as a dumping ground for excess parts, with the apparent impact of reducing operational or project costs. For the company, this is a false economy.

Example

For a particular project a special type of electrical cable was ordered. The project only required 100m of cable, but the minimum delivery was 1,000m. The excess 900m was transferred to inventory. The total cost of the cable was $200,000.

As the project used only 10% of the cable, the project was charged $20,000. An 'investment' in inventory of $180,000 was made, despite there being no further need for that cable. The project manager believed that he had saved the company money because he got the cable for his project for only $20,000.

The company had, however, spent $200,000. This expenditure was not immediately recognized because it did not appear in the profit and loss statement or project budget. Had the project been made to recognize the entire cost of the cable, then perhaps a different solution would have been found.

These Myths Have A Dual Impact

Achieving sustainable inventory reduction relies upon the implementation of new management practices, measures, and reporting to drive new behavior. As in most areas of management, however, there are truisms that often prevent action, or worse, give the appearance of action but no sustainable benefit. These are called the 5 Myths of Inventory Reduction.

These myths have the dual impact of adding to the inventory investment and preventing action to achieve sustainable reduction. Overcoming these myths requires a universal recognition of the cash impact of inventory and an understanding of the behavioral issues that impact management decisions. Only after these myths are recognized and overcome can true inventory optimization be achieved.

Why You Might Hold More Inventory Than You Really Need

In all materials and spares inventory management systems, excessive inventory is created by a combination of systemic and structural issues and changes in circumstance. In fact, at least twelve major reasons why you might hold more inventory than you really need have been identified. Three of these reasons relate to change that may not have been recognized and nine relate to systemic and structural issues in your business.

Change-Related Issues

There are three main issues that relate to change:

1. The initial max-min is based on a best guess, which is not revisited unless a stockout occurs.

2. Improvements in supply chain efficiencies have not flowed onto inventory holdings.

3. It is assumed that you must make the inventory investment and little effort has gone into minimizing that investment.

In each of these cases, an initial inventory stocking decision was made based on the best understanding of the situation at that time. Perhaps with little or no information or history, assumptions had to be made about supply, demand, and the reorder point requirements. The supply chain capability may have been quite limited or inflexible, or suppliers may not have been willing to support their business through consignment stock.

However, it is entirely possible that the original circumstance has now changed and that the original decisions have not been revisited. With the busy-ness of managing materials, spares, and operations, these decisions are unlikely to have been revisited if the changes have not resulted in a stockout, which of course would be a trigger for action.

In the period of time since the original parameters were set, it is likely that:

- You now have a better understanding of the demand requirements.
- You have worked to improve the supply chain or your suppliers have.
- The commercial environment has changed so that suppliers may be prepared to invest to support their ongoing relationship with your business.

In reviewing the opportunities for inventory reduction, it is necessary to ensure that the issues relating to these types of change are understood. This must include both where change has occurred, but has not yet been recognized, and where you could change the future result through positive action today.

Systemic and Structural Issues

The systemic/structural causes of excess inventory can be harder to influence than the change-related causes, but should be understood none the less.

1. Misaligned Responsibilities

   Based on years of working with companies in different industries and all over the world, it is my opinion that the misalignment of responsibilities is the number one reason why companies hold more inventory than they really need. The level of inventory that is held is almost always a function of who is responsible for determining what should be held and who is responsible for the level of investment. Fundamentally, these are two very different considerations and form the basis of the inventory tension discussed in Chapter 1.

   If one person has the authority to add an item to inventory, but is not responsible for the financial impact on the company of that decision, then it is likely that the inventory will be overstocked. This overstocking is not driven by any malice or poor intent. It is driven by basic human nature that seeks to maximize our personally measured outcomes. For example, managers responsible for sales and gross margin are unlikely to be motivated to minimize inventory if they believe that more inventory gives them more opportunity to make more sales. Similarly, engineers responsible for keeping a factory producing are unlikely to be motivated to reduce inventory if they are not held accountable for the dollar value invested and if they believe that they can minimize their own risk of equipment failure by stocking up on inventory.

   It is by aligning both the responsibilities and the authorities relating to materials and spares inventory management that an appropriate focus on true inventory optimization can be maintained.

2. Management Systems

   The way that companies are set up to process the issue of materials and spares from their storeroom or warehouse, and place orders to replenish materials and spares, can have a significant impact on inventory holdings. Unfortunately, many companies fail to recognize the impact that their own systems have on their overall materials and spares inventory management. They tend to think only about the external supply chain and its implications. By including a review of their own internal processes, they can make significant gains in reducing the replenishment cycle.

When an item is sold or used, how long is it until the inventory system recognizes the movement of the item and takes corrective action? A slow system will require an increase in the quantity of safety stock required, as the ROP calculations are based on the entire lead time or replenishment cycle. Examples where long cycles are particularly obvious include where there is a significant period between replenishment reviews or where there is an overly long reorder process.

For example, one company that operated without a computerized system reviewed its reorder requirements once a week (on Wednesday mornings). This weekly approach meant that any item that reached a reorder point on Wednesday afternoon would not be reviewed for a full seven days. If requirements were reviewed and reordered daily, the company would save a full week's stock holding. If they operated with an average holding of 10 weeks stock, they would drive a 10% reduction in inventory!

Another company, which was computerized and could essentially review their needs instantly (if they so chose), had a purchasing process that often took a week or more before the actual order was placed. Again, streamlining the ordering process would save that company five or six days of stock holding and a significant level of its investment.

A third company was slow to receipt their deliveries into the storeroom and enter them into the computer. This delay also adds to the restocking lead time and increases the inventory held.

3. The Number and Location of Stores
   The more locations that exist for storing an item, the greater the quantity will be stored. The generation of greater quantities from having more locations is primarily driven by the perceived need for safety stock at each location. Obviously, then, if there are fewer locations, there will be less inventory. The savings that can be generated in safety stock is the main driver behind companies centralizing warehousing and inventory holdings (in addition to the real estate costs!).

As a rule of thumb, the safety stock requirements increase with the square root of the number of stores. So, an organization with two inventory stores will hold 41% more safety stock than if they had only one. (The square root of 2 is 1.41). If they had three stores they would hold 73% more safety stock; four stores 100% more safety stock, and so on.

In addition to the number of inventory stores, the location of those stores also systematically drives inventory levels. A location in a remote rural area is likely to need relatively higher levels of safety stock than a location in a major city. The additional safety stock reflects the likely proximity to suppliers and the extended lead-time for supply. Great care should be taken to select an inventory location in order to get the best balance of customer service, inventory cost, and logistics costs.

4. High Safety Stock Levels
Previously it was discussed that changes may occur in the supply or demand for a product that would enable a change in safety stock holdings. The key there was to be aware of the change and make the necessary adjustments.

In some circumstances though, safety stock levels are set across a range of products on a rule-of-thumb basis. Setting safety stock in this way assumes that all items exhibit the same attributes when it comes to inventory requirements. These are usually the attributes of the item requiring the greatest level of inventory and so will systematically overstock all the other items. Setting safety stock levels with a broad brush approach is another example of the problems with a 'one size fits all' approach to inventory management. Getting the most out of your inventory investment requires inventory setting to be done on an item-by-item basis.

5. Economic Order Quantity
Previously this chapter discussed the 5 Myths of Inventory Reduction, including the myth that Economic Order Quantities save money. The central argument here is that economic order quantities are only economic if enough of the product is used or sold.

In addition, order quantities are sometimes described as being economic, but are really just for convenience. For example, orders may be placed for three or six months stock because it saves having to place another order in that period. However, in the current age of technology and administrative automation, ordering items for the sake of convenience rather than to minimize investment would seem to be out of place.

Today there is automated ordering, electronic delivery acceptance, and electronic funds transfer for payment. If the volume of traffic through these systems were doubled, the costs may increase marginally due to server or bandwidth issues, but they would be far outweighed by the savings from an inventory reduction.

6. Unrecognized Obsolescence
   Accounting standards require that once an item is recognized as obsolete, its value is written down to zero. The problem, of course, is that the item needs to be recognized as obsolete first. It is not unusual for an item to become obsolete, but not be recognized as such.

This is because most inventory systems drive people to manage active stock; thus, obsolete stock is likely to go unnoticed until or unless it appears on an aged item or non-moving stock report. However, items do not get into these reports until it is too late to proactively manage obsolete stock.

In some cases, companies operate reward systems that effectively 'punish' people for accumulating obsolete stock. This has the unwanted consequence of inadvertently encouraging inventory managers not to recognize obsolete stock. The intent is well meaning in that it is supposed to discourage purchasing items that become obsolete. However, in practice, these systems don't allow the admission of a mistake or for the natural occurrence that some items are likely to become obsolete. Managing obsolescence in this way drives a behavior to not recognize obsolescence as the recognition impacts the individual's recognition or rewards.

7. Product and Range Complexity
   Imagine if you ran a business with only one stock item and that item only came in black. Then, your inventory management would be simple and (hopefully) your holding would be minimal.

   Holding a range of items means that the inventory holding will by necessity increase because each item will have batch and safety stock requirements. The increase in inventory will occur whether the complexity is a wide range of different products or spares or a wide range of options on a similar product or spares. For example, a hardware store will hold a wide range of product to attract customers; a wire manufacturer will hold a wide range of sizes and tensile strengths of wire. This also occurs with maintenance materials and spares inventory when a company has a wide range of different machines.

   Once it is recognized that product or range complexity is a key reason for the level of inventory, efforts can be focused to minimize the complexity. Actions might include rationalizing the range, increasing scrutiny on individual products rather than applying a 'one size fits all' policy, or standardizing equipment.

8. Project and Special Inventory
   Previously it was discussed that one of the main reasons for holding inventory is to act as a temporary measure to accumulate stock prior to major projects. Holding inventory in this way occurs particularly in businesses with seasonal activities, rolling projects, major sales periods, maintenance shutdowns, and so on.

   These events mean that they regularly hold higher-than-standard levels of stock in anticipation of usage in a particular time frame. Often people say 'Our inventory would be fine if it weren't for these events.' The fact is that these events are likely to be a key part of your business and, rather than be an excuse for holding excess stock, they should be a target for closer management to ensure that the investment is minimized.

   Managing these events efficiently requires an approach to match supply more closely with demand and a method of clearing stock that is not used in the 'window.' That is, inventory that is not used for the purpose for which it was purchased should be disposed of. Typically people say, 'We might use it later' and then never do. Companies should bite the bullet and take care of the excess inventory sooner rather than later.

9. Operating Requirements of Supply
   Almost all manufacturing and supply systems operate at some level on batching quantities. This could be the minimum quantity for a machine set up and production run, the minimum quantity for efficient shipping/transport, or the minimum quantity for the time and effort involved in actually purchasing the item.

   Having minimum inventory levels that are based on supply rather than demand almost invariably means that inventory will be held at some point in the supply chain. One way to minimize the quantity of inventory that is affected by this is to minimize the batches that are required at the bottleneck that causes the inventory to be held. In many situations, making changes to bottlenecks will be difficult and could be expensive if it requires major equipment upgrades.

   However, operating efficiently within your existing system requires an approach that ensures operating to the absolute limits of the existing bottlenecks. To operate to the limits of the existing bottleneck, you need to understand whether the bottleneck is real, imagined, or a result of tradition ('We always do it that way!'). Breakthroughs occur when changes are made not only to eliminate these bottlenecks, but also to minimize their impact by working the system to its limits rather than accepting the problem.

Chapter 7

The Spare Parts Storeroom

The Central Hub

The spare parts storeroom is the central hub of materials and spares management activity. It is, of course, where the spares are stored (unless you allow squirrel stores). It is where they are supplied from when needed, and where they are delivered to when ordered. The management of the storeroom and the process therein can have a significant impact on your materials and engineering spares outcomes. Think about the impact of a storeroom that:

- Is slow to receive goods into the stores system.
- Is poorly laid out, making it difficult to find items.
- Is poorly labeled, making it difficult to identify items.
- Exerts control over holding levels without reference to their customers, the end users.
- Is slow to reorder items.
- Doesn't seek to protect the items in the store from environmental effects.
- Makes it difficult to record items removed 'after hours'.
- Doesn't inspect inwards goods to ensure quality.
- Doesn't conduct cycle counts to ensure accuracy.
- Doesn't maintain its catalogue to ensure visibility of items.

The result of these actions (or inactions) is a materials and spares management system that fails to deliver the right parts, in the right place, at the right time, for the right reason. And this is just a selection of the problems that can be found in storerooms all over the world!

Although each of the above is important, most can be dealt with quite readily when they are identified by responsible management. This section of the book is dedicated to addressing some issues that are less obvious and require some difficult action to correct. The so-called 'squirrel stores' are a cultural and behavioral issue that is not addressed solely by changing process. Lean Thinking has been applied across many areas of manufacturing, but does not yet appear to have made its way into the storeroom. The idea of organizing the storeroom to suit the customer is not implemented as widely as it probably should — in fact maintenance is sometimes seen as the enemy of the storeroom rather than the customer. Addressing these more difficult issues will take you further along the path of achieving and implementing smart inventory solutions.

Squirrel Stores and Why You Would Be Nuts to Keep Them

Breaking the locks was the only option. It was 2am and Line 1 had stopped completely. The good news was that we knew exactly what the problem was and how to fix it. We also knew that the spare part we needed had been in the storeroom earlier in the day — I had seen it there myself. The bad news was that it was no longer there and we didn't know who had taken it or where they put it. We were pretty sure that one of the dayshift crew had taken it and put it in his locker. Waiting was not an option so locks had to be broken. We just hoped that we found the part before doing too much damage.

Sound familiar? This scene is played out in maintenance workshops all over the world. Maintenance team members take parts and put them away in their own stores and sometimes, when really needed, the part cannot be found. The team members do this either because they think it is convenient or that it saves time. Convenient and time saving for them, but what about the rest of us!

Let's face it, reliability and maintenance people are different. They have a unique position in the world. We all know that when things go wrong, maintenance gets the blame. But when things go right, production gets the credit. As a result, maintenance people hoard spare parts, like squirrels keeping nuts for the winter. That's why these unofficial stores are often referred to as 'squirrel stores'. Look around almost any workshop and you will find spare parts that are being held in squirrel stores, 'just in case'.

The problem with this, as demonstrated above, is that when parts are held outside of the official storeroom or inventory management system, they actually impact the rest of your inventory holding for that part. Not only in the obvious ways of poor availability and access, but also in less obvious ways relating to inventory levels, operational expenditure, and even your reliability program — more on that in a moment.

First, let's understand why these stores exist. The main reason is trust. That is, trust that your official store will have the required parts when they are needed. If your storeroom management is unreliable, this erodes trust in the system. Also, if team members know that other team members are squirreling away parts, then they might do the same — just in case. No one wants to be caught short. Not only does it let the plant down, but it is personally inconvenient.

More than just being inconvenient, not having the spare part can be a real hassle. If the plant is down at 2:00 am and it is your job to fix it, but there is no spare, then you get the hassle from production — even though it is not your fault. Most people would believe that it is better to avoid all those problems and keep your own little emergency squirrel store — again, just in case.

A third reason is a personal rationalization that squirrel stores improve service (or at least reduce downtime) by reducing the time needed to go and get the spare from the official store. Squirrel stores are usually held closer to the plant (or at least closer to the team member) than the official store; hence, the time to access the store is reduced.

No matter what the logic or reason, squirrel stores are ultimately a cultural issue and they need to be managed on that basis. This requires building trust in the system, communicating the negative impact of squirreling, modeling and encouraging the right behavior, and not allowing any exceptions.

Now, how do squirrel stores really impact your inventory levels, operational expenditure, and reliability program? Also, why would you be nuts to allow your team to keep squirrel stores? Here are six reasons:

1. You will hold more inventory.

Duplicating the parts being held in your official store by holding parts in a squirrel store obviously adds to your inventory. However, it is the flow on effects of squirrel stores that can be much, much worse. You might be surprised to realize that in addition to duplicating your inventory, squirrel stores can also significantly increase the level of spares held in your official store. How? Through a mechanism that I call Induced Demand Volatility (IDV).

IDV occurs when your team takes more spares than are actually required so that they can put some into their squirrel store. This behavior produces false data on usage and shows higher volatility than is really the case. This higher volatility then results in a need to hold more safety stock — after all the reason that you hold safety stock is to account for volatility. The following section 'How Much Do Squirrel Store Really Cost?' shows a situation where induced demand volatility could increase spares holdings by 264%!

2. You will spend more money.

Obviously, the parts in the squirrel store and the extra parts in the official store have to be paid for. Therefore, much more cash is tied up than would otherwise be the case. What many people don't consider is that this situation diverts funds from other and more useful purposes. Still waiting for that tool to make your life easier? Perhaps the money that you need is tied up in your squirrel store!

3. You will spend more on your operating budget and skew your reporting.

When your team removes more items from the store than they really need at that time, the costs have to be charged somewhere. Guess where — one of your operating budgets! Not only does this limit your ability to manage and improve your reliability (with what will already be a tight or underfunded budget), but also it skews your reporting of costs by bringing forward costs that you could have incurred later. In many cases you may even be paying for parts that never get used, which leads to the next point.

4. You will have increased obsolescence.

Is anyone really keeping track of those squirrel stores? Of course not. So, you have spent the money and when the item eventually becomes obsolete (as everything does), the squirrel stores will contain items that should have been used or should not even have been purchased! The only time they will be cleaned out is when someone decides to tidy up their squirrel store or workshop and you know that they will then just throw the parts in the trash.

5. You will increase your downtime.
This is perhaps the worst part of the squirrel stores phenomenon. If the 'unofficial' parts are held in a locker or tool kit so that only the 'owner' can access them, then the rest of your team cannot access them. If you have a breakdown and need that part right away, you might not be able to get to it or might not even know that it is there! The irony here is that the part was being held in order to improve service and the approach actually made things worse. The result of this scenario is an increase in 'official' holdings, increasing expenditure even further.

6. Your reliability program will be endangered.
As mentioned previously, when your team keeps squirrel stores, they skew the data on usage. But this doesn't just impact your expenditure. It also means that your official records will show higher demand than actual at some times and lower demand than actual at others. If you are trying to perform any sort of analysis to understand your failure patterns, this data will be useless at best and, at worst, misleading. All that money spent on reliability training, software, gadgets, and cultural change could be wasted because of a failure to control squirrel stores.

Unfortunately squirrel stores are almost a fixture of maintenance departments. They result from the mindset of reliability and maintenance professionals who are passionate about reducing downtime and take equipment failure personally. This drives them to hoard items that they can use later and to short cut the system to try to improve response times. However, this approach does not work. Squirrel stores are a blight in your system. They can have a significant and detrimental impact on your expenditure and your reliability program. You would be nuts to allow or endorse them.

How Much Do Squirrel Stores Really Cost?

The following example demonstrates the inventory effect of squirrel stores. For this example, let's consider a part that is used weekly and therefore has an average demand of 1 unit per week. This type of part is a major target for squirrel stores as holding them reduces the number of trips to the storeroom.

Let's compare two situations:

1. No Squirrel Store: The item is removed from the storeroom as needed — 1 per week.
2. Squirrel Store: The item is removed 2 at a time with movement every 2 weeks.

The demand data for these two situations is shown in Table 7-1 and the demand profile for these two different situations is shown in Figures 7-1 and 7-2.

	Demand Data	
Week	No Squirrel Store	With Squirrel Store
1	1	2
2	1	0
3	1	2
4	1	0
5	1	2
6	1	0
7	1	2
And so on…		

Table 7-1: Comparison of Demand Data

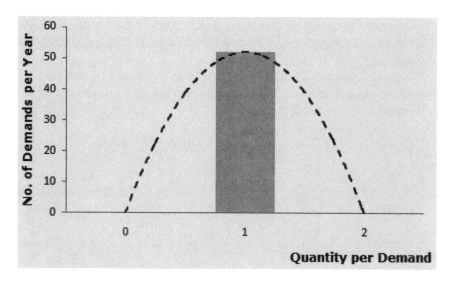

Figure 7-1: Demand Profile Without a Squirrel Store

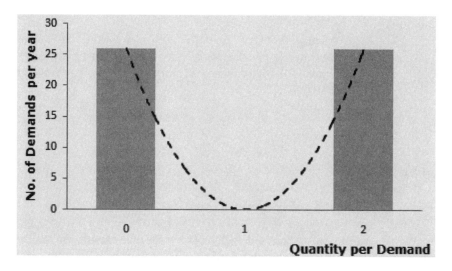

Figure 7-2: Demand Profile With a Squirrel Store

It is clear from Figures 7-1 and 7-2 that, while in each case the average is one demand per week, the demand profile is not just different, it is completely opposite.

A common way to calculate the inventory needs in this situation is by using a Gaussian distribution. This approach is familiar to most people as it can be represented by the formula:

Reorder Point = (Usage rate x lead time) + safety stock

Alternatively:

ROP = (D x LT) + csf x MAD x $\sqrt{(LT)}$

Where,

RP   = reorder point

D   = average demand per week (for our example this is 1 per week)

LT   = Lead time in weeks (let's assume 4 weeks)

csf   = customer service factor (or availability factor) (here we will use a csf of 2.56; this assumes a 98% availability)

MAD   = Mean Average Deviation — a measure of demand variation. In this example with no squirrel store, this measure is 0 (there is no variation); with the squirrel store the MAD is 1.

$\sqrt{}$ = square root

Results

Scenario 1: Using the above formula and data.

Most people are surprised when they see the result of holding inventory in a squirrel store — the Reorder Point in your official store can be more than double the Reorder Point without the squirrel store. Table 7-2 shows the results for this scenario.

Measure	No Squirrel Store	With Squirrel Store
Re Order Point	4	10
Average Inventory	2.5	9.1

Table 7-2: Results for Scenario 1

This result then means that the average level of inventory held in your official store, if you allow a squirrel store, is 264% greater than the average holding without the squirrel store (see Figure 7-3). This is not due to the items held in the squirrel store but due to the Induced Demand Volatility (IDV) that the squirrel store creates in your official store. The IDV changes the calculation of safety stock in the above formula, leading you to hold too much inventory.

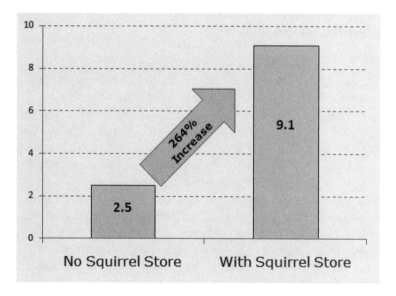

Figure 7-3: Average Inventory Increases 264% with a Squirrel Store

Scenario 2: Override the calculation and manually set your reorder point to 4 for both scenarios.

Let's now assume that you understand the impact of the IDV on your calculation and decide to manually set the reorder level for both situations to 4, knowing that you only ever use 4 items during the lead time for supply. In this scenario, the average for the squirrel inventory holding reduces to 3.5 items (including the items held in the squirrel store).

This is still 40% higher than the situation without the squirrel store!

Do you still think that squirrel stores don't cost much?

MRO and Engineering Spares: Lean, Mean, or Somewhere In Between?

Lean Manufacturing (or 'Lean' as it has become widely known) gained worldwide recognition after being acknowledged as the philosophy that helped make the Japanese auto industry the industrial powerhouse that it has been for the past several decades. The publication of books such as Toyota Production System, The Machine That Changed the World, and Lean Thinking helped make the philosophy accessible to tens or even hundreds of thousands of companies worldwide.

Although originally developed as a way of embedding higher quality and lower costs in manufacturing organizations, Lean principles have since been applied in many different organizations and management situations. One of these is the Engineering Materials and Spares Storeroom.

One of the key benefits of becoming Lean in a manufacturing organization is savings in inventory and working capital. Therefore, it is perhaps ironic that many organizations applying Lean to their storeroom activities end up being mean rather than lean. Similar to the earlier discussion on JIT, lean techniques were developed to manage a production environment. By applying the immediate and linear thinking associated with production line management to the timelines and complexity of the storeroom, companies can miss real opportunities to embed quality and reduce costs in their storeroom activities. In an effort to appear efficient, the methods used in Lean are 'cherry picked' and applied without adopting the whole philosophy. This approach can result in the achievement of working capital (or financial) targets while holding too much of the wrong inventory and too little of the right inventory. This really is 'mean' thinking!

The key to applying Lean is described in the forward to the book Toyota Production System by Taiichi Ohno:

> *To understand its tremendous success, one has to grasp the philosophy behind it without being sidetracked by particular aspects of the system...if the system is introduced without being part of a total philosophy... problems will ensue.'*
>
> Muramatsu Rintaro
> Faculty of Science and Engineering
> Waseda University

According to Taiichi Ohno, the philosophy behind lean is based on the absolute elimination of waste. This philosophy is embodied in the 7 Wastes identified in Toyota Production System. Before moving on to explain these and how they can be applied to the Engineering Materials and Spares Storeroom, we must first understand the principles of lean.

The Principles of Lean

Developing a lean storeroom involves much more that applying the process review of value stream mapping or adopting Kan Ban style techniques. It is about applying the principles that support the concept of lean.

In the late 1980s, a team from MIT undertook a major research project to investigate and understand different manufacturing techniques for automobiles. For this study, the authors visited 90 automotive assembly plants in 17 countries. This research resulted in the book The Machine That Changed the World. This book documented the effect of Lean Production. However, it was only when the authors wrote the book Lean Thinking that they clarified the principles for relentlessly eliminating waste.

The five key principles and their application to MRO and engineering spares are:

1. Specify Value: Value must be as defined by the customer.
   The MRO and engineering spares storeroom has, in effect, two customers: the users of the spares and the financers of the spares. The users specify value as having the right spare available, at the right time. The users only care about spare parts sitting on the shelf to the extent that this helps assure them that the parts will be available when required. Having parts on the shelf adds no real value to the user — only its availability when required adds value. This is a subtle yet important difference.

   Similarly, the financier sees value as only financing the minimum required value of spares to operate the plant. This means having the right spares there for the right reason. Underinvesting in the right spares adds no value to the financier because it impacts plant operation.

   So the user wants the right spare available at the right time and the financier wants the right spares stocked for the right reason. The needs of the user and financier are often seen as being in conflict, but they are actually far more aligned than many people realize.

2. Identify the Value Stream: Define the value adding steps and eliminate the waste.
   Managing engineering materials and spares is a complex task primarily because of the number of spares held and the fact that they may all have different attributes relating to supply and demand. Couple this with the physical activity of managing a storeroom and there is plenty of room both to misunderstand the value stream and for waste to occur. Examples of the 7 Wastes will be covered shortly.

3. Flow: Eliminate bottle necks and keep the product moving.
   This is the most obvious aspect of lean thinking and it relates as much to keeping spare parts, people, and information moving through the system as it does to keeping manufactured parts moving in a manufacturing system. However, when spares are sitting on the shelf, they are not flowing. Having spares sitting on shelves for long periods might be hard to change because of the nature of some spares, such as

insurance spares. This idea can just as readily be applied to the people and information involved with the management of the inventory.

4. Pull: Base activity on the demand from the next step.
   This principle is central to lean thinking and the application of this principle to spares management requires the matching of supply with use. At first glance the existence of a random failure pattern (as oppose to production demand patterns) in spares management makes this principle difficult to implement. However, by working with both suppliers and the users, it is not impossible. It does, however, require teamwork.

5. Perfection: Pursue perfection through continuous improvement.
   MRO and engineering spares have previously in this book been referred to as the 'forgotten investment'. Companies often decide their holding requirements in a 'set and forget' manner —this means that the spares holdings are not reviewed unless a stockout occurs. Similarly, the practices associated with supply and fulfillment are not reviewed unless there is an external driver such as a system upgrade. As a result, little continuous improvement in stock holding is achieved, making it difficult to implement Lean Thinking.

The application of Lean requires that these five principles are applied in a systematic and continuous way. This is shown schematically in Figure 7-4.

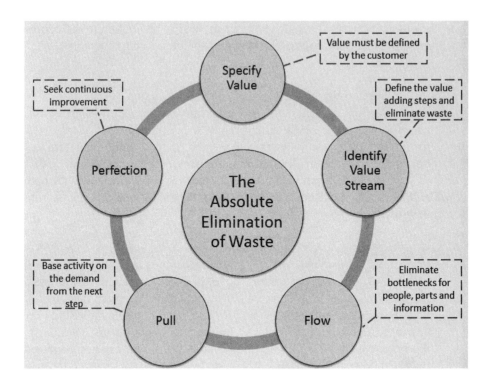

Figure 7-4: The Principles of Lean Thinking

Applying the 7 Wastes to MRO and Engineering Spare Parts

The Toyota Production System is based on the idea of building in quality and reducing costs through the absolute elimination of waste. To support this system, Toyota identified 7 Wastes to be eliminated.

Value Stream Mapping is the technique that is most widely used to identify opportunities for eliminating waste. It can be applied to store room operations. However, for spare parts there are some complicating factors. In Toyota Production System, Taiichi Ohno says that efficiency must be improved (and waste eliminated) at each step; this is a complication for spare parts.

For example, consider a storeroom with several thousand spare parts, each of which may have a different supply and demand pattern. In this case, the steps being mapped will be many and varied. Applying value stream mapping to each part is possible but unrealistic. A more pragmatic approach is to apply value stream mapping to the physical and informational storeroom activities and utilize the 7 Wastes more generally for reviewing the actual supply and management of spares.

Other complications include the dichotomy between the philosophy of just in time (JIT) and the need for slow moving and insurance spares, as well as the mismatch that often occurs between demand or usage quantities and minimum order quantities when commercial terms are dictated by the supplier. These complications don't prevent the opportunity to identify and eliminate waste by applying Toyota's 7 Wastes. They just require a perspective that understands their unique attributes. The following is provided as a thought starter on how waste occurs and can be eliminated in spares management.

1. Waste Through Mistakes

   One of the main insights of lean is that most costs are assigned when a product and/or a process is designed. With respect to spares, this translates to the decisions that are made during the Plan and Set Parameters stage and then included in the storeroom (see Figure 7-1, The Materials and Inventory Management Cycle). This is the design point for spares management, the point at which decisions are made that affect the spares holding for years to come. Yet often these 'Day one decisions' are paid too little attention. Many subsequent spares issues could actually be prevented through greater focus at this time.

   Other mistakes that result in spare parts waste include: poor parts selection, having the wrong parts in stock, and paperwork/keying errors that result in inaccurate data.

2. Waste in Process

   Traditionally this applies to process issues such as set-up time and excessive processing. In spare parts management, this translates to excessive bureaucracy and excessive process steps to create, order, and manage spares. Often these issues are created by a desire for control, a lack of teamwork in determining appropriate approaches, and inadequate policy and procedures. Eliminating this waste requires giving people the skills and knowledge required so that they may efficiently complete their tasks without undue supervision or control.

3. Waste in Over Producing

   There are two types of 'over production' in spares management. One is over ordering of stock, which usually occurs when economic order quantities are pursued inappropriately. It must be recognized that economic quantities are not always economic, especially if the full quantity ordered is not used in a timely manner or is ultimately written off (see Chapter 6, The Five Myths of Inventory Reduction). This type of waste can be eliminated by implementing appropriate stocking policies to provide ordering guidelines.

   The second example of 'over production' occurs when users remove more items from the storeroom than they really need. Sometimes this happens for 'just in case' reasons; often it is a matter of wanting to have their own spares supply in their own tool kit. In either case, this action results in the over ordering of spares and creates inaccuracies in data, both in usage levels and in volatility, and renders software-based optimization useless. Eliminating this waste requires a cultural shift and the development of trust that the system will provide the availability of spares as required.

4. Waste Through Storage

   Lean management seeks to eliminate storage buffers; this is one of the outcomes from using a Kan Ban system. Spares management invariably relies on inventory buffers and it is unlikely that these will be eliminated in all cases. However, setting safety stocks too high creates a bigger buffer than necessary. This bigger buffer is a waste that can be eliminated.

   In addition, many tradespeople have the habit of creating their own spares squirrel stores, thereby duplicating holdings. This habit creates another waste that can be eliminated. As mentioned above, eliminating this waste requires a shift in culture rather than a shift in process.

5. Waste in Transportation

   Good storeroom management requires that storerooms are laid out in a manner ensuring that the high demand items are located closer to the entry/exit (and closer to the floor in high rise layouts) than the slow moving items. This approach to layout of the storeroom eliminates much of the travel for personnel in storing and retrieving items and is a clear elimination of waste. (See also the next section, Is Your Storeroom a Pit Stop or a Supermarket?)

6. Waste in Time Spent Waiting

The timely processing of the receipt of goods into the storeroom can have a significant impact on the level of inventory held and the visibility of that inventory to the user. Companies that are tardy in this area (usually on the basis that 'it is there if it is needed') fail to realize that this practice adds to their effective lead time (increasing safety stock), Furthermore, if users check the stock on hand using a computer, they will not see the items waiting to be processed. Tolerating a situation where the users must visit the store to check holdings, because they know that items aren't processed, embeds the waste in the system. This waste can be eliminated by applying appropriate process discipline and receipting goods in a timely manner.

7. Wasted Movements

The ultimate wasted movement in spares management is the selection, cataloguing, ordering, receiving, storing, and stocktaking of items that are not ever used. This doesn't apply to insurance or critical items that are held in the actual hope that they are not ever used. It does apply, however, to those items that are ordered without sufficient forethought or planning — items that eventually are written off and scrapped.

Another wasted movement occurs when too many items are removed for (say) a preventive maintenance task and then some items are returned to the store. This results in extra work for everyone. That is not to say that the return to store is a waste. After all, the alternative is stock held outside the store system. But waste is created when the initial selection of too many items happens in lieu of proper planning. In both of these examples, the waste can be eliminated by appropriate forethought and planning.

Correct Application Enables Efficient Operations

The documentation of the Toyota Production System and the subsequent development of lean thinking have, for more than 30 years, enabled companies in many different industries to eliminate waste and improve their operating outcomes. The principles of lean are well understood. However, their application to MRO and engineering spare parts inventory is often hampered by the linear thinking associated with manufacturing processes. This has led to instances where the inventory holding is more mean than lean.

Correctly applying lean principles to maintenance store rooms and MRO and engineering spares enables more efficient operations. It also helps to eliminate excess and unnecessary spares holdings, resulting in greater stock turns and improved availability.

Is Your Storeroom a Pit Stop or a Supermarket?

Ever worked in or visited a facility with a very large spare parts storeroom? While walking through and understanding how they operate, you can sometimes actually feel like being in a local low-cost suburban supermarket rather than in the storeroom of a high tech manufacturing facility. By that I mean that there may be long aisles that are not clearly marked, high use materials not always near the user's entry/exit, and boxes in the aisles waiting to be unpacked. To make the picture complete, you may even see a maintenance guy walking up and down an aisle saying, 'Where do they keep the pump gaskets?'

Now compare this scenario with the cool efficiency of the pit crews in the Formula 1 racing when a car comes in for fuel, tires, and so on. Of course a pit crew has it relatively easy because they know exactly what is required at each visit. Still, it is interesting to think about how the large industrial storeroom (and maybe even your own storeroom) could be improved and made more efficient. As with almost all things in operations, the starting point is to focus on the customer. In this case, think of Maintenance as the Storeroom's customer. Storerooms exist to enable spares to be available to the maintenance team so that they can do their work. It is reasonable to expect that the customer will want:

- Parts available as per the agreed stocking policy.
- Parts in the places that the system says that they are.
- Easy identification of parts and their stock codes (assuming that this must be noted when the part is removed).
- Quick and easy retrieval of parts.

Most people understand the concepts and relationships between layout and workflow so I won't repeat them here. I will say, however, that that one key to implementing these concepts is to ensure that the layout works for the customer — that is, the maintenance team. Therefore, if your storeroom allows direct entry by the maintenance team, then the layout should work for them. Even if you do not allow direct access, the above issues apply because this is what minimizes the time spent by the maintenance team while waiting for parts to be retrieved.

Too often storerooms are set up for the convenience of the storeroom personnel. The workflow is set up based on their receipt and issue of parts, not to maximize the customer outcome. Sometimes it is appropriate to sacrifice a little of the storeroom time efficiency if it then improves the time efficiency for the maintenance crew.

A couple of other techniques and their potential issues are:

- Storing items for one type of plant or machine together can seem like a good idea. But it also can mean that both high usage and low usage items are stored together. Mixing items in this way can then lead to storing other high usage items in a less optimal location — the optimal location is taken up by storing the low usage items of a type of plant or machine together with the high usage items for that plant of machine.

- Storing all like items together (e.g., bearings) also seems like a good idea. But doing so can result in confusion in parts and numbers, incorrect stocking (similar parts in the wrong bin), problems with returns to store if they are not controlled by the store personnel, and, of course, the co-location of high and low usage items, as mentioned above.

So what is your storeroom: an F1 pit stop or a supermarket? The closer you can get to being a pit stop, the less down time you will have and the more productive your maintenance team will be.

Chapter 8

The 7 Actions for Inventory Reduction

The Bathtub Principle

Before moving on to explain the 7 Actions for Inventory Reduction, it is important to understand them in the context of the 'bathtub principle'. To effect a change in the water level in a bathtub, either up or down, you need to change either the input or the output, that is, either adjust the faucet or the drain. With materials and spares inventory management, the issue is very similar. To effect a reduction in your inventory level, you need to either increase the output or reduce the input. This is what is known as the Bathtub Principle.

Applying this principle to materials and spares inventory management means that the actions you can take will fall into one of two categories:

1. Take more items out – remove items from inventory
2. Put fewer items in – don't put items into the inventory

Recognizing the Bathtub Principle enables you to cut through all of the jargon and over-complication that gets written and discussed as part of inventory management. At a fundamental level there are only two strategies for inventory reduction; all actions are variations of those strategies. This is a key concept because many people find inventory reduction to be a complex exercise involving supply chain review and initiatives, IT installation, and major operational change. At the heart of that though, you are only trying to find ways to take more items out, or put fewer items in. The Bathtub Principle is demonstrated in Figure 8-1.

Once you understand that this is the mechanism that is required to actually make change happen, you can take all decisions in that context. Ask yourself, 'Will this action take more items out or put fewer items in?' Simple as that.

Figure 8-1: The Bathtub Principle

Although the Bath Tub Principle is simple and guides the 7 Actions for Inventory Reduction, two other scientific approaches are used to ensure that the actions are mutually exclusive and completely exhaustive (sometimes called MECE). These are Hypothesis Driven Analysis and Double Loop Learning.

Hypothesis Driven Analysis

Hypothesis Driven Analysis has been the basis of scientific development for more than 200 years. It involves developing hypotheses about the subject and then either proving or disproving the hypotheses. The hypotheses that are proven then form the basis for ongoing development. To develop the 7 Actions for Inventory Reduction, a scientific approach was applied to a management discipline—inventory reduction.

Double Loop Learning

Double Loop Learning was developed in the 1970s by Harvard Professor Chris Argyris, who recognized that in almost any type of problem solving people work within constraints that may or may not be real. He called this approach to problem solving Single Loop Learning.

There are two problems with Single Loop Learning. First, you can never improve beyond your self-imposed constraints (Argyris calls these governing variables). Second, you may not realize that you are imposing these constraints on your thinking (otherwise they are unlikely to be constraints). Achieving breakthrough improvements in any field requires a challenge to the constraints inherent in the original thinking. Professor Argyris called this Double Loop Learning. Figure 8-2 shows how Double Loop Learning extends the thinking of Single Loop Learning.

When developing both the Inventory Cash Release® Process and the 7 Actions for Inventory Reduction, we first challenged the constraints inherent in most inventory management thinking and then created hypotheses that were focused on the cash rather than the SKU.

As a result, the 7 Actions for Inventory Reduction encompass all of the alternatives that can be addressed to achieve inventory reduction. They are 100% complete. They are the only 7 actions that can be taken. Thus the Inventory Cash Release® Process is simultaneously both the simplest and the most complete approach to inventory reduction. Figure 8-3 lists the 7 Actions for Inventory Reduction; the Inventory Cash Release® Process is described in Chapter 9.

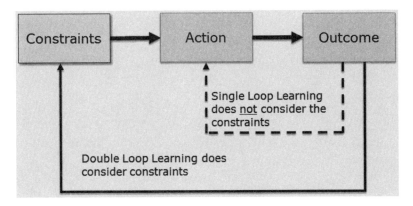

Figure 8-2: Double Loop Vs. Single Loop Learning

Take More Items Out

Action [#]1: Have someone else hold it and/or pay for it

Action [#]2: Sell excess and obsolete stock

Action [#]3: Eliminate duplication

Put Fewer Items In

Action [#]4: Change the factors that drive safety stock

Action [#]5: Reduce reorder stock

Action [#]6: More closely match delivery with usage

Action [#]7: Reduce value of items held

Figure 8-3: The 7 Actions for Inventory Reduction

Action #1: Have Someone Else Hold It and/or Pay For It

The best way to reduce the investment in inventory is to eliminate the investment altogether. Reducing inventory in this way can be achieved without risk and with full access to the inventory by having someone else hold and/or pay for the inventory. This approach, sometimes known as *consignment stocking*, transfers both the ownership and the management of replenishment of the inventory to the vendors. Your company only pays for what it uses, when it uses it.

By entering into a consignment stock arrangement, the investment in inventory is shifted to the vendors. They are also responsible for managing orders and replenishment, generally with the requirement that they meet an agreed service level. This service level might be an availability level or it could be that they never have less than x number of units in stock.

Yet consignment stock is often thought to be too expensive. (See Chapter 6: The Five Myths of Inventory Reduction.) The theory is that the vendors will, in time, seek to recover the cost of holding the stock through price increases. But assuming that there is still competition for your business, if a price increase happened, you could still switch vendors or move out of the consignment arrangement. Setting up a consignment arrangement does require careful contract development.

A slightly less effective approach is called Vendor Managed Inventory (VMI). With VMI, you pay for the inventory upfront, but transfer replenishment and management of the inventory to the vendors. VMI gives the vendors greater visibility of stock usage which, in turn, means that they should then be able to better manage the supply chain to minimize your investment in inventory. Figure 8-4 compares the characteristics of consignment stock and vendor managed inventory.

One of the key benefits to the vendor of both consignment and vendor managed arrangements is that they provide full visibility of the supply chain through to your customers and users. Rather than seeing only your order, which will cover requirements for some period (which could be a month or several months), they will be able to see actual user demand. This extra visibility will enable them to better time their production and supply chain activities to produce a better result for both parties.

Another key benefit for the vendors is sole supplier status. This status guarantees your business until you decide to change the arrangement.

Action #1: Have someone else hold it and/or pay for it

This approach reduces or eliminates investment in inventory while maintaining operational needs.

Consignment Stock
- ✓ Transfers inventory from vendor's warehouse to your site
- ✓ Effectively reduces the number of stock holding points

Vendor Managed Inventory
- ✓ Vendor takes responsibility for holding stock on our site
- ✓ Also reduces the number of stock holding points

These approaches enable improved supply efficiencies through improved visibility of both supply and demand.

Figure 8-4: Action #1

In applying these arrangements, be sure to maximize your buying power. Larger companies that are made up of a number of separate and independent operations often do not take advantage of their buying power as a group. In some cases, one part of the company may have negotiated a consignment arrangement, but not passed this information onto the other parts of the company.

Setting up a consignment arrangement usually takes considerable effort. Input is required from the vendor and your own purchasing /procurement team and legal and finance departments. Some of the key issues that need to be resolved include:

- When does ownership change?
- Who takes the risk associated with storage?
- Who takes the risk of obsolescence?
- Who pays for the cost of funding the stock?

- What checking procedures are to be used?

Resolving these issues requires a cooperative approach from both parties. Consignment stocking should be a win–win situation. Demanding that vendors make a sudden and significant investment to replace all of your stock is likely to be met with an argument relating to cost. But seeking to run down your existing stocks and bringing the vendors on-line gradually enables them to spread their investment over time. They can also gain an understanding of the demand issues and fine tune their approach accordingly.

There is considerable effort involved in setting up consignment and vendor managed arrangements that work, but the rewards are just as great. As the example below shows, a well-structured consignment arrangement can deliver a 100% stock reduction and ongoing operational savings.

Implementation Checklist 8-1 shows the actions to consider with the implementation of Action [#]1.

How to Recognize a Good Consignment Deal

One of the key questions that people ask me at workshops and seminars is: 'What does a good consignment deal look like?' Of course, a consignment deal will result in a shift of inventory ownership from your company to the supplier's; otherwise, it is not a consignment deal. The deal would be preferable if there was a price reduction, but this is not an absolute requirement for a good deal. In fact, there probably are no absolutes for a good deal because each company and supplier relationship is different. However, here are three issues to consider and to help decide if a consignment deal is a good one:

1. Change in the Supply Chain.
   A good consignment deal should result in some changes in the supply chain dynamics that provide an overall cost benefit. Otherwise, all the deal does is shift the inventory cost and risk from one party to another. Suppliers may accept a shift without cost savings in order to be certain about winning your business. But in the medium term, they are just as likely to resent the shift and try to claw back some costs through price increases.

Changes in the supply chain can be many and varied. Here are some examples:

- Manufacturing efficiencies: Because suppliers are no longer responding to your purchase orders (which may be erratic in terms of timing), they may be better able to schedule the production or internal ordering of the components they supply. This will reduce their manufacturing costs.

- Transport efficiencies: Again, because they are not responding to your purchase orders, suppliers may be able to opportunistically use transport options. For example, they may have a half load going close to your location and choose to fill the truck to make the delivery of both loads more efficient.

- Order management efficiencies: Not having to process a continuing stream of purchase orders, raise occasional invoices, and so on saves expense in these areas.

- Inventory efficiencies: Often a supplier will hold stock for a major customer while the customer also holds stock; this combination doubles up on the inventory that is held. A good consignment deal might be able to reduce the overall inventory that is jointly held.

2. Risk Management
   The number one fear that companies have with consignment deals is that the supplier will not hold sufficient stock for their needs. Therefore, if a stockout occurs, they are left stuck without the item they need when someone else was controlling the inventory. Managing the situation where someone else manages the inventory requires appropriate monitoring systems to review stock levels and associated risk. The key here is to realize that consignment does not mean that you, as the consignee, give up all responsibility — you need to put in place risk management processes for monitoring stock levels and ensuring that appropriate quantities are held. This is no different than how you might review the holdings of your own inventory.

3. Financial Control Mechanisms

   Aligned with but different from risk management is the requirement for financial control mechanisms — that is, the way that stock usage is recorded and invoices raised. A good consignment deal will make this process seamless with your own internal systems so that 1) there is little or no extra burden and 2) there is confidence that the charges reflect actual usage.

Example of Action #1

Most examples of consignment stock are straightforward. A deal is done, transition effected, and a 100% reduction in inventory results.

In a recent case, though, there was an extra benefit. The supplier took advantage of the new visibility of demand and its opportunity to produce and deliver to its own timetable. They also used the customer's premises as a storage location (there was plenty of room) and occasionally drew down on the stock that was stored there.

As a result, the supplier was able to cut it costs and subsequently its price to the company by 4%. The company held $2M of inventory and used $4M per year of the product. As a result of the arrangement, they not only reduced their inventory investment by $2M, they also benefited from an ongoing cost reduction of $160,000 per year.

Implementation Checklist 8-1

Action #1: Have Someone Else Hold and/or Pay For It

☐ Identify current and potential suppliers.

☐ Check if similar consignment deals exist elsewhere in the company.

☐ Discuss opportunity with purchasing department.

☐ Check current contractual arrangements.

☐ Identify potential bundling for consignment.

☐ Contact suppliers and request quote.

☐ Collate supplier responses.

☐ Select preferred supplier.

☐ Develop changeover plan and timetable.

☐ Advise store(s) of changeover plan and timetable.

☐ Implement planned actions.

☐ Ensure IT adjustments made.

☐ Monitor changeover.

☐ Confirm final handover of inventory.

Action #2: Sell Excess and Obsolete Stock

Another great way to eliminate the investment in inventory is simply to sell off items that are excess or obsolete. Operationally, excess and obsolete items generate little interest because they do not create emergencies. As discussed previously, the focus tends to be on items that stock out or, at the least, need reordering. Items that are excess or obsolete do not fit either of these categories so only tend to get reviewed if there is a review program undertaken. Excess and obsolete items add no value to your business and so should be eliminated. Figure 8-5 shows diagrammatically how excess and obsolete stocks relate to your value adding inventory.

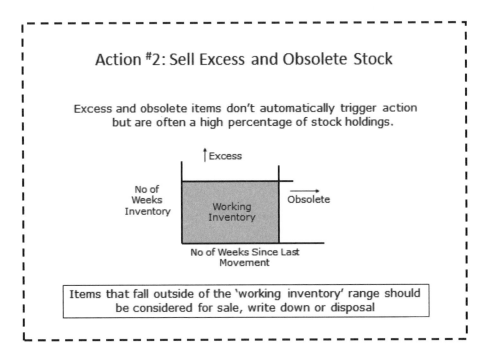

Figure 8-5: Action #2

For obsolete items, there are accounting standards that require that any item that becomes obsolete be written down when it is recognized as being obsolete. In many cases, though, items become obsolete, but this is not recognized because there is no systematic approach to identify them. If they become obsolete and you don't recognize this, they add to the value of your inventory, but don't add value to your business. Also, because accounting standards write down excess and write off obsolete inventory, it is tempting and easy to leave these items on 'auto pilot' and not manage them proactively. Although it may seem that the problem will just go away, the reality is that the company has spent its cash on these items and received nothing in return. In effect, this is a complete waste of money! Looking at excess and obsolete inventory as a potential waste of money completely changes the perspective of managing inventory.

For those times when you come across items that have not yet been confirmed as obsolete, but which are suspected to be identified as obsolete, a new term has been identified — probsolete. Probsolete means that you think that the item is obsolete, but that it needs further investigation, that is, probably obsolete. This term helps categorize the items without placing an accounting constraint until obsolescence is confirmed.

In effecting an inventory reduction, there is an argument that says removing written down items from inventory adds no value as the item is already written down. There are, however, four key reasons why removing obsolete items is important.

1. It ensures a discipline that you don't just bury your mistakes by making them obsolete. You can learn from the items that become obsolete and perhaps change your future buying patterns to minimize future obsolescence.

2. It costs money to store, count, and sometimes maintain obsolete items. The fewer items you have, the lower your inventory management costs.

3. If you can sell the item, you can actually retrieve some value. The value may be minimal compared to the original or book cost, but on a cash basis, every dollar of income is better than zero sitting in inventory.

4. There may be tax gains from removing obsolete stock. In some countries, a company that removes obsolete stock can gain a tax offset equal to the value of the stock multiplied by the company tax rate. The tax offset is usually only available when the item is physically

removed from the company's premises and ownership. On a cash basis, this tax break is equivalent to receiving revenue of the same value. Please note that this observation is no substitute for accounting advice; you should take separate advice from your accountant before planning in taking up a tax offset.

Excess items occur for a number of reasons. You may have simply ordered too much. There may be returns that came in after restocking. Or, you may have applied Action [#]5, adjusted the safety stock level, and created excess stock. Generally people are slow to react to excess stock because there is a belief that the items will get used or sold eventually. The key question is, when is eventually?

Although a standard definition of excess is 'stock held beyond the set maximum holding,' it may or may not make sense to sell off that stock if it can be used in a suitable period. Defining what is suitable will be a matter of specific company circumstance. In determining a suitable period, it is important to realize two things:

First, the longer the time frame, the greater the risk of deterioration or obsolescence, in which case, the value may deteriorate rapidly.

Second, holding onto an item costs money. At a minimum, stockholding costs about 10% of the stock value per year. However, as discussed previously, it can be argued that holding stock costs more than 20% per year. Using the lower value of 10%, if excess stock of $100,000 is held for three years, the cost could be $30,000! This is a cost that will not be seen in the operating budget but it is, none the less, a cost to the company.

Adopting an approach to manage both excess and obsolete stock in a rigorous and timely manner occasionally results in a need to make difficult decisions to remove or scrap excess or obsolete stock. However, in the long term, adopting this approach is likely to result in reduced obsolescence and lower holding costs.

Implementation Checklist 8-2 shows the actions to consider with the implementation of Action [#]2.

Example of Action [#]2

This company operated a well-known ERP system and the inventory managers were convinced that they had good control and little to gain. But they were only partially right. They did have good control over the items that moved regularly; however, they still gained significant benefits from Action [#]2.

The company had $7.5M in inventory and from this they identified:

- $400,000 in obsolete items not previously provisioned
- $350,000 in items no longer required by them but that could be sold or transferred to another business in their group
- $1,400,000 in items that were overstocked and could then be preferentially used, limiting the expense of purchasing other items

This is a total $2,100,000 in reductions, equal to 28% of the inventory holding!

Implementation Checklist 8-2

Action #2: Sell Excess and Obsolete Stock

☐ Review if item is in excess.

☐ Review if item may be obsolete.

☐ Confirm excess or obsolete status and quantity.

☐ Discuss with purchasing if they can assist with sale.

☐ Identify potential buyers.

☐ Negotiate deal and select buyer.

☐ Discuss disposal of obsolete stock with Finance Department.

☐ Develop disposal plan and timetable.

☐ Advise store(s) of disposal plan and timetable.

☐ Implement planned actions.

☐ Ensure IT adjustments made.

☐ Monitor disposal.

☐ Confirm final disposal of inventory.

Action #3: Eliminate Duplication

The third action relating to 'taking items out' is to eliminate duplication. By definition, duplicated items add no value and so should be a prime candidate for inventory reduction. There are a number of ways that items can be duplicated in an inventory system. Within a single store, the same or similar items may be held as different item numbers. Across a network of stores, the same or similar items may be duplicated when a shared access might be preferable.

Examples of duplication include:

- Specifying new SKUs that are only marginally different from those in stock.

- Holding the same item specified by different equipment suppliers using their own inventory codes. This is particularly the case where the OEM supplies the inventory.

- Holding hold stock at two (or more) locations with safety stock at each.

- Holding safety stock when your supplier holds safety stock as well.

Despite the advantages of computerization, inventories that consist of thousands of SKUs can be unwieldy to manage. It is unlikely that anyone knows everything that is held. It is also possible that more than one person is making recommendations on what ought to be held. With this degree of complexity, it is not just possible; it is likely that duplication will occur. The message here is simple: seek to identify and eliminate duplication.

The issue with duplication is not just that the item is held twice; it is that duplication increases the safety stock that is held. (Recall that the safety stock requirements increase by the square root of the number of holding points.) Therefore, eliminating duplication doesn't eliminate the duplicated inventory; it eliminates the safety stock that results from the duplication.

Previously it was discussed that reducing the number of holding points helps reduce safety stock. Action #3 is similar. However, in this case, the duplication may not be across a network, but within a single store!

The advantage of eliminating duplication is not just limited to safety stock. Eliminating items helps simplify the range to be stored, counted, and managed. It reduces the inventory management costs.

The impact of eliminating duplication is also highly dependent upon the variability of demand and the circumstance in which the stock is held. If the items turn over quickly and have reordering parameters set appropriately, the gain may be small. If the items are slow moving, savings can be significant.

Consider this extreme example. If two items both turn over 10 per week and each has overnight replenishment, it is likely that almost no safety stock will be held. If the items are combined, the one item now turns over 20 items a week and still has overnight replenishment. Again almost no safety stock would be held. The main advantage here is in simplification of management and the reduction in procurement administration.

However, if the items are slow moving, say one demand per quarter, but still have a short replenishment time, there will be opportunities for inventory reductions. For example, if two items meet the above profile but are held separately, it is likely that a minimum of one of each would be held. By combining the items, it would be acceptable to hold only one in stock, a saving of 50% in inventory investment. The effect of this reduction if the item has a high dollar value is obvious.

One tip on reviewing items for duplication is to use the supplier's representatives to review your stock. No supplier would want to miss an opportunity to eliminate a competitor's stock and substitute their own, so they will usually review your stock for you for free. Just make sure you select your preferred supplier!

Implementation Checklist 8-3 shows the actions to consider with the implementation of Action [#]3.

Example of Action #3

Using the slow-moving example from above, let's assume that items A and B are identical, but held as separate items from separate suppliers in the same inventory.

Item A is typically used or sold at a rate of 1 per month, has a safety stock of 1, a reorder quantity of 3, and a replenishment time of 1 day. Item B is typically used or sold at a rate of 2 per month, has a safety stock of 1, also a reorder quantity of 3, and a replenishment time of 1 day.

As two separate items, the average holding of A and B will be 5 items (Safety plus half reorder stock—this equation is explained more fully under Action #5).

If rationalized to one item, there will now be usage of 3 per month and could probably still have safety stock of 1, and a minimum reorder of 3. The average holding would now be 2.5 items.

This is a 50% reduction in average inventory.

Implementation Checklist 8-3

Action #3: Eliminate Duplication

☐ Check if item is held across other stores.

☐ Check if supplier holds significant stocks locally.

☐ Review if it is possible to adjust internal or external holdings.

☐ Confirm duplication to eliminate.

☐ Confirm location(s) to hold stock.

☐ Develop change over plan and timetable.

☐ Advise store(s) of change over plan and timetable.

☐ Implement planned actions.

☐ Ensure IT adjustments made.

☐ Monitor changeover.

☐ Confirm final handover of inventory.

Action # 4: Change the Factors That Drive Safety Stock

Action #4 is the first of the actions relating to 'put fewer items in.'

It was discussed previously that safety stock is used to act as the buffer between supply and demand. Figure 8-6 represents a series of replenishment cycles for a product. In this figure, the shaded area represents the safety stock. The horizontal line represents the reorder point.

In this example, the safety stock is set so that, with the expected lead-time and demand, the stock usage reaches the safety stock level at the time of replenishment. But what happens when demand is greater than expected or lead-time longer than expected? As shown, the inventory level goes below the safety stock.

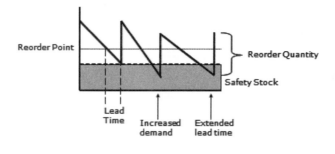

Figure 8-6: Action #4

This example demonstrates that safety stock has a definite purpose and must be managed to ensure that it fulfills its purpose. There are ways, however, of reducing the investment in safety stock without impacting the availability of stock or the ability to provide the required buffer. In fact, there are five ways to reduce safety stock:

1. Increase the speed and reliability of replenishment.
    In the example shown in Figure 8-6, you can see that extending the lead-time for replenishment results in using safety stock. The replenishment time and the variation in that time partially drive the quantity of safety stock required.

    By taking actions to shorten both the replenishment time and the variation in replenishment time, you can hold less safety stock. These actions could include reviewing your own processes (remember the example where the order took a week to process) or working with suppliers on improving their response. Increasing the speed and reliability of replenishment has formed the basis of most supply chain improvements in the past ten years. The key difference here is that you know exactly the value of the SKU you are seeking to reduce and can easily determine the viability of making any change and at what point it does or does not make sense.

2. Smooth the demand pattern.
    The other driver of safety stock is demand fluctuation. In the example shown above, we saw that an increase in demand resulted in a usage of safety stock. By taking actions to smooth the demand pattern, you can safely reduce the level of safety stock. This is particularly relevant in situations where demand is driven by artificial factors such as month end accounting and terms of business or, for engineering spares, the usage of items in a factory. The example at the end of this section demonstrates both these circumstances.

3. Hold only the right amount of stock.
    In earlier chapters it was discussed that inventory that does not stock out rarely gets attention for review. Similarly, it is also likely that safety stock levels have been based on an initial estimate of

requirements. If that estimate was high and the item never stocks out, it is unlikely that safety stock would be revisited.

One way to minimize the safety stock is simply to review the required holding based on historical data once sufficient time has passed to provide a sound basis for review (being sure to check the validity of your data). Depending upon the volatility of the product, collecting the required data could take anywhere from 6–12 months or longer.

4. Identify changed circumstances.
Supply chain improvement has been the buzz word of industry since the early 1990s. Even if your company has not made major advances in supply chain management, it is entirely likely that your suppliers have. They may have a vastly improved capability, but you are not taking advantage of their good work. Of course, if your company has made advances in supply chain management, make sure that you take advantage of that also.

It is also possible that demand patterns have changed. In engineering, there may have been major maintenance improvements aimed at increasing reliability, but with the flow-on effect of enabling a reduction in spares holdings. For direct inventory items such as raw materials and finished goods, the demand may have slowed or changed in characteristic, meaning different demands on your plant and equipment.

By reviewing the current circumstance for an item you can reset the holding parameters and reduce inventory holdings.

5. Have fewer stock holding points.
It has been mentioned a few times that there is a rule of thumb that safety stock increases with the square root of the number of stocking points. For example, if there are two stocking points, 41% more inventory will be required than if there was one. For three points, 73%. In theory, reducing from five stocking points to three would enable a 22% reduction in safety stock.

Review the locations at which inventory is held and question whether the number of holding points is still logical.

Implementation Checklist 8-4 shows the actions to consider with the implementation of Action #4.

Example of Action #4

Modifying Demand for Consumable Spares

In this case, the item was a high usage consumable used in the manufacturing process. This particular consumable was used in a regular and predictable fashion with almost constant and even demand.

The tradesmen who use the item always sign out two or more weeks of stock at any one time as a way of minimizing the number of visits to the central store. They do this despite the fact that they visit the store on an almost daily basis. Signing out the item in this way had the impact of creating artificial demand spikes for an item with flat and predictable usage. This in turn resulted in increasing the quantity that the firm held at any one time (although much of that was expensed when booked out of the store) and increased safety stock due to the demand spikes.

By applying a limit of, say, two days usage to be booked out at any one time, the company would not only reduce the total quantity on hand, it would also be able to substantially reduce its safety stock.

Implementation Checklist 8-4

Action #4: Change The Factors That Drive Safety Stock

☐ Identify the real level of safety stock for this item.

☐ Identify the drivers of safety stock for this item.

☐ Identify the mechanism by which safety stock could be reduced.

☐ Engage internal personnel responsible safety stock need.

☐ Contact suppliers responsible for driving safety stock levels.

☐ Confirm changes that can be made.

☐ Identify the viability and risk impact of reduction.

☐ Develop change over plan and timetable.

☐ Advise store(s) of change over plan and timetable.

☐ Implement planned actions.

☐ Ensure IT adjustments made.

☐ Monitor changeover.

☐ Confirm reduction of inventory.

Action #5: Reduce Reorder Stock

We have seen in Action #4 that the safety stock level has a major impact on stock holdings. However, this is only part of the equation. The other key factor is the reorder quantity. In the classic saw tooth representation, the average quantity of stock held across a period of time is a function of both the safety stock level and the reorder quantity. Action #5 focuses on the reorder quantity to reduce the average stock holding.

In Figure 8-7, the shaded area once again represents safety stock. On the left side of Figure 8-7, there is a situation similar to that shown for Action #4. On the right side of Figure 8-7, the reorder quantity is reduced by half. This change has had the effect not only of reducing the average stock holding but also increasing the required frequency of delivery. Assuming that the delivery increase is not onerous, then changing the inventory ordering in this way is a very effective way to quickly reduce stock holdings with no increase in risk. The horizontal lines represent the average stock holding under each scenario.

With this classic saw tooth pattern, it is a simple mathematical calculation to determine the average stock holding. That is, average stock equals the safety stock plus half the reorder quantity.

The reason that there is no increase in risk is that Action #5 works on the cycle stock part of the Gaussian equation. Chapter 2 discussed calculating the ROP for a normal distribution using the formula:

ROP = (Demand x Lead Time) + Safety Stock

Action #5 does not change any of the elements of this calculation. Therefore, your risk has not changed.

Recall from the beginning of this chapter that one of the only two strategies that can apply to inventory reduction is to put fewer items in. The point at which you can act on the decision to put fewer items in is when you reorder stock. Therefore, **this is the point in the cycle at which you have greatest influence over the stock holding for the next cycle**. By reducing reorder quantities, you will significantly impact stock holding.

It is also worth considering the logic applied when using a maximum and minimum approach to setting levels of stock holding. Many companies set a minimum based on their level of comfort at meeting expected demand with that level of stock. The minimum is based on demand characteristics.

They then manage their stock holding to their maximum stock setting, that is, the level of stock they are comfortable holding based on supply characteristics. Managing their stock holding in this way drives their stock holdings to the maximum.

Why not change the approach and manage to the minimum? The minimum is the level that will meet demand and also minimizes the investment in inventory.

Implementation Checklist 8-5 shows the actions to consider with the implementation of Action [#]5.

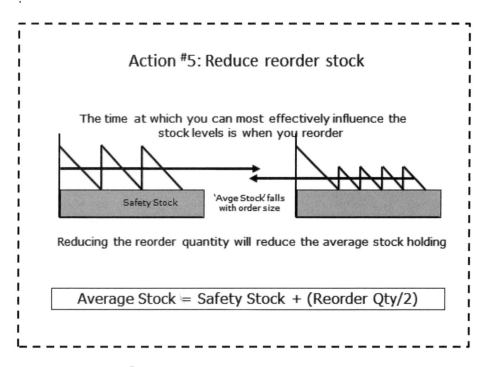

Figure 8-7: Action [#]5

Examples of Action #5

The practical impact of Action #5 action is dependent upon the relative level of safety stock compared to operating stock. Two examples are given to show the relative impact.

<u>Low Safety Stock</u>

This company orders 12 weeks supply of an SKU that takes one week to be restocked. Let's assume that they hold two weeks supply as their safety stock.

The average stock   = safety stock + (ROQ/2)
                           = 2 + (12/2) = 8 weeks stock

Now, assume that they change the reorder quantity to 4 weeks and reorder every month.

The average is now = 2 + (4/2) = 4 weeks stock

This simple change produces a 50% reduction in the average holding of that SKU.

<u>High Safety Stock</u>

This company orders 4 weeks supply of an SKU that takes one week to deliver. In this case, the demand is highly volatile and they hold 4 weeks supply as safety stock.

The average is now = 4 + (4/2) = 6 weeks stock

Now assume that they change their reorder quantity to 2 weeks and order every fortnight but still hold the same level of safety stock.

The average is now = 4 + (2/2) = 5 weeks stock.

Even in this more extreme example, the average stock reduction is 16%!

Implementation Checklist 8-5

Action #5: Reduce Reorder Stock

☐ Review current reorder quantity and frequency.

☐ Identify probable causes/reasons for current quantity and frequency.

☐ Contact suppliers to discuss impact of changing quantity and frequency.

☐ Confirm changes that can be made.

☐ Develop change over plan and timetable.

☐ Advise store(s) of change over plan and timetable.

☐ Implement planned actions.

☐ Ensure IT adjustments made.

☐ Monitor changeover.

☐ Confirm reduction of inventory.

Action #6: More Closely Match Delivery with Usage

Each of the actions discussed so far has been looking at ways to reduce the physical number of items held. However, one of the variables that drive the cost of holding inventory is how long you hold the inventory. This represents the amount of time that you might have to finance the working capital required to buy the inventory (sometimes referred to as the time value of money).

Reducing the time that you hold inventory has the same impact as putting fewer items in. This is because if you can shorten the length of time in which you hold the inventory, you can significantly reduce the average inventory holding and directly impact the holding cost. This is shown diagrammatically in Figure 8-8.

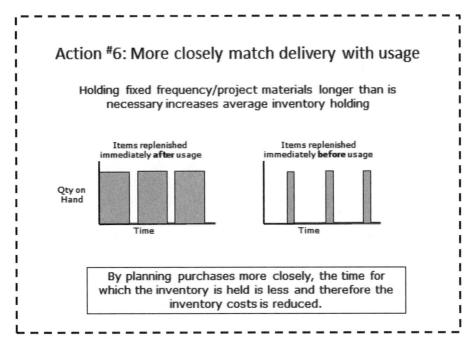

Figure 8-8: Action #6

In many cases it may be hard enough to predict how much of an item you need, let alone get the timing of delivery right. After all, you hold inventory as a buffer between supply and demand. There are some cases, however, where you can make decisions about timing without impacting the risk. These are cases where you have project stock, regular frequency use items in engineering, or sale/special event stock in wholesale of parts.

In each of these cases, the usage/requirement date is reasonably well known. By planning the supply of that inventory close to the event, the average inventory holding is reduced substantially.

Figure 8-8 represents a fixed frequency of usage. On the left, the item is restocked immediately after use. Here you can see that the stock then waits until the next usage before going quickly through a usage/restock cycle.

On the right of Figure 8-8, the same item is restocked just prior to being used. Here there is no stock for long periods when the item is not required. You can see from the amount of white space on the right that the average holding will be very much lower. The example on the next page quantifies this.

This philosophy is a central part of a JIT approach to inventory management and often is thought to need sophisticated MRP controls or Kan Ban systems. The approach can be applied, though, with just a little forethought on the inventory needs and the requirements of demand.

One warning: Some companies operate reservations systems whereby stock can be held for a particular use or customer. If these systems are applied indiscriminately or don't recognize the acceptable level of risk, they can cause overstocking. For example, assume that an item will only be reordered when it reaches a minimum quantity of 5. This means that it has been previously determined that 5 items are sufficient to cover the restocking cycle. If there are 10 in stock and someone reserves 5, a poorly tuned reservation system will order a restock even though the minimum has not been physically reached. Reordering the stock in this way would result in stock being brought in weeks ahead of when it will be required. If the full quantity of reserved stock is not used (as often happens), then the item will be overstocked and you will have invested in inventory that you do not yet need.

Poorly tuned reservations systems are a common issue in manufacturing and production environments.

Implementation Checklist 8-6 shows the actions to consider with the implementation of Action [#]6.

Example of Action #6

A company replaces a wearing part in its plant every 12 weeks. The part is held as a stores item in order to eliminate any need to actually plan usage.

Assume that the item costs $1000 and takes just one week to be restocked.

If the items is restocked immediately after the part in stock is used (the pattern on the left of Figure 8-8), the part will be in stock for 11 out of 12 weeks or approximately 47 weeks of the year. At a conservative holding cost of 10%, this would cost the company $90.00 per year to hold. [(47/52) x $1000 x 10%]

If the item is ordered to be restocked, one week before being required, (the pattern on the right of Figure 8-8) the item will be in stock for 1 out of 12 weeks. The holding cost will now be $8.33 per year, a reduction of 90%!

Now translate that saving to the thousands of items that might benefit from this approach and the saving will be massive.

Implementation Checklist 8-6

Action #6: More Closely Match Delivery With Usage

☐ Identify the triggers for demand of the item.

☐ Determine if triggers can be reliably predicted.

☐ Engage internal personnel responsible for identifying need.

☐ Contact suppliers and determine minimum lead time for supply.

☐ Confirm that changes can be made.

☐ Develop change over plan and timetable.

☐ Advise store(s) of change over plan and timetable.

☐ Implement planned actions.

☐ Ensure IT adjustments made.

☐ Monitor changeover.

☐ Confirm reduction of inventory.

Action #7: Reduce the Value of Items Held

Reducing the cost of items held is the last of the 7 Actions and is probably the most obvious, but perhaps because of that the most overlooked. Most cost reduction actions are aimed at reducing the cost of an item for the obvious profit and loss impact. However, cost reductions also have an impact on working capital and the investment required for inventory.

The other six actions discussed in this book are aimed at reducing the quantity held in stock or the time for which it is held. Action #7 works on reducing the purchase cost and, therefore, the overall value held.

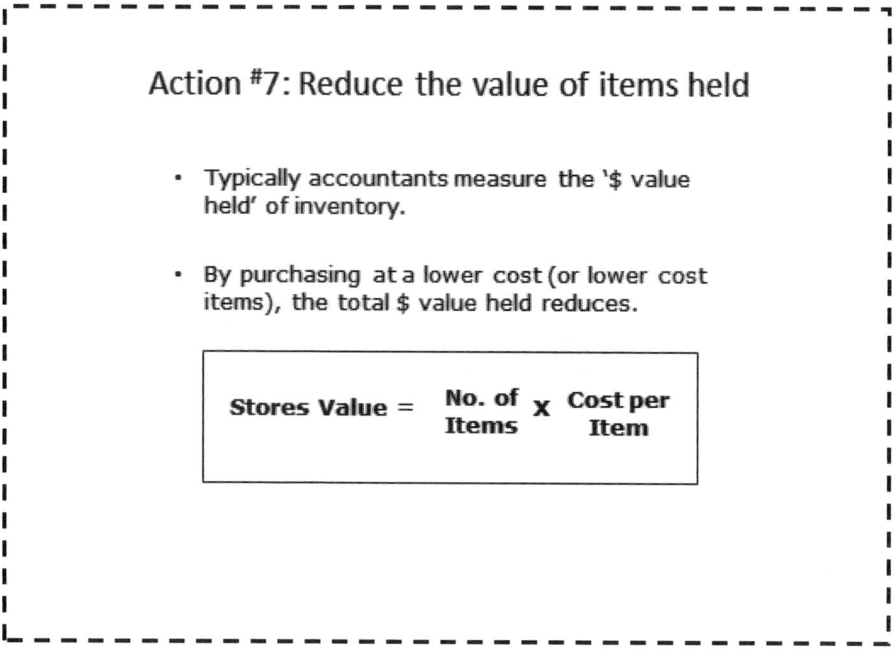

Action #7: Reduce the value of items held

- Typically accountants measure the '$ value held' of inventory.

- By purchasing at a lower cost (or lower cost items), the total $ value held reduces.

$$\text{Stores Value} = \begin{array}{c}\text{No. of}\\\text{Items}\end{array} \times \begin{array}{c}\text{Cost per}\\\text{Item}\end{array}$$

Figure 8-9: Action #7

There is no reason why the approach would be different from a normal cost reduction exercise — it's just that the initial starting point is different. The target may not even be the actual unit price, but could be the delivery costs. For example, can delivery be consolidated without impacting inventory holdings? Do you pay for fast delivery when slower delivery will do?

Here are some of the approaches that can be applied to review the purchase cost of inventory:

- Volume concentration—across sites or with suppliers
- Product specification review
- Joint process improvement
- Bundling
- Unbundling
- Expanding the supplier base to increase competition
- Concentrating the supplier base to drive greater surety with vendors

Or just plain old renegotiate!

Action $^{\#}7$ is not advocating a comprehensive strategic sourcing review, but a targeted price review aimed at specific SKUs that are high value held items. From this perspective, this action becomes a very manageable task to undertake.

Be wary, however, of trading off inventory holding for cost reductions, that is, buying more of an item to get a cost reduction. As we have seen in the 5 Myths of Inventory Reduction, this approach may not actually provide an economic benefit. There needs to be a clear economic benefit for any change.

Implementation Checklist 8-7 shows the actions to consider with the implementation of Action $^{\#}7$.

Implementation Checklist 8-7

Action #7: Reduce Value of Items Held

☐ Engage with purchasing for pricing review.

☐ Identify current and potential suppliers.

☐ Identify mechanism by which price could be reduced.

☐ Determine economic viability and impact on inventory reduction.

☐ Contact suppliers and request quote.

☐ Collate supplier responses.

☐ Select preferred supplier.

☐ Develop changeover plan and timetable.

☐ Advise store(s) of changeover plan and timetable.

☐ Implement planned actions.

☐ Ensure IT adjustments made.

☐ Monitor changeover.

☐ Confirm reduction in inventory value.

Chapter 9

Inventory Process Optimization™

Beware Implicit Assumptions

When presented with the need to review inventory, many organizations immediately seek out one of the many inventory optimization software programs on the market. They do this with the belief that optimization is the ultimate outcome. Although the belief is correct, the problem is that the optimization software doesn't really optimize your inventory.

Optimization is one of the most overused words in management today. But what is optimization? The Concise Oxford Dictionary defines optimization as 'the most favorable condition; the best compromise between opposing tendencies; best or most favorable.'

Using the above definition of optimization, it can be said that typical optimization software does calculate the 'compromise between the opposing tendencies of cost and availability.' However, the outcome is achieved only by recalculating the required holding and safety stock based solely on historical data.

The fact that optimization software programs base their calculations on hard data such as usage history makes the approach particularly appealing to some people. The claim is usually made that with this kind of solid input, the results must be right.

But this is not the case.

Historical data, no matter how accurate and clean it is, only tells us what has been. When it comes to your inventory investment, you should be more interested in what could be. What software does not do, and cannot do, is consider the alternatives. Software does not and cannot recognize whether the outcome you have experienced is a result of your own behavior or whether something has changed that you can take advantage of.

Using only the historical data approach forces you to make assumptions about the characteristics of both demand and supply for your inventory. It does not force you to challenge those assumptions. Even though many people don't realize it, the historical data approach forces you to assume that:

1) What happened in the past will happen in the future.
2) You cannot change these outcomes or the behaviors influencing them.

The approach used by the so-called optimization software forces you to work within constraints that may or may not be real. It is an example of Single Loop Learning. This is why the software optimization approach doesn't truly optimize. It almost never involves consideration of the broader issues that influence your inventory holdings. It does not utilize all seven actions for inventory reduction and does not constitute double loop learning. At best, it only recalculates within a set of assumed constraints; it doesn't challenge those constraints.

There is one other failing of the software optimization approach. It can be misleadingly inefficient.

With a few simple keystrokes, software can analyze thousands of inventory items in just a few minutes. Although the attraction of this may seem obvious, the key question is: What happens next? The all-encompassing approach of software analysis is indiscriminate when it is applied to the entire inventory just because it can be. The broad brush approach seems efficient because everything gets reviewed. However, what this also means is that your team then spends their time and energy reviewing and deliberating over inventory items that may have little or no impact on your inventory reduction target. This because the software treats all inventory the same — no matter what its value. You do not gain the benefits of 'maximum result, minimum effort' that is achieved through the Inventory Process Optimization™ Method.

The Inventory Process Optimization™ Method

This book has explored many aspects of engineering materials and spares inventory management, including the mechanics of inventory management, the MIM cycle, the financial impact of inventory, the influence of people, processes, policies and procedures, issues and myths, the operation of the storeroom, and the 7 Actions for Inventory Reduction. Now we need a framework in which to apply this knowledge and understanding — this is the Inventory Process Optimization™ Method.

The Inventory Process Optimization™ Method combines knowledge of parts usage, procurement, and supply chain issues with a review of behaviors and the management processes that drive them. The method has been specifically developed for application with MRO, engineering spares, and other indirect inventory and materials management. It recognizes and understands the impact of people on inventory outcomes. Therefore, its application involves training a wide range of personnel and influencers to ensure a consistent understanding is developed by those who will ultimately drive the results.

To achieve this, the method is divided into two streams of activity: an Individual Parts Review and a Management Process Review. These two streams are summarized in the following pages and then explained in detail after that. The application of the two streams is shown in Figure 9-1.

Stream 1: Individual Parts Review

Stream 1 is an Individual Parts Review using the Inventory Cash Release® Process and it comprises three key elements.

1. **Focus on the high dollar value items.**
   In every single review of MRO and engineering materials and spares inventory that I have seen, the inventory value adheres to the 80:20 rule. That is that 80% of the value is tied up in 20% of the inventory items. This means that reductions in working capital can be achieved by reviewing a small percentage of the items in stock, saving a lot of time, energy, and money in conducting a review and implementing improvements. Software solutions will usually include the entire inventory and so at least 80% of the effort is wasted on items that will have little or no impact on your problem.

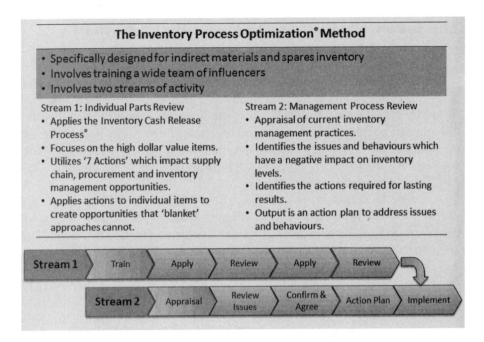

The Inventory Process Optimization® Method

- Specifically designed for indirect materials and spares inventory
- Involves training a wide team of influencers
- Involves two streams of activity

Stream 1: Individual Parts Review
- Applies the Inventory Cash Release Process®
- Focuses on the high dollar value items.
- Utilizes '7 Actions' which impact supply chain, procurement and inventory management opportunities.
- Applies actions to individual items to create opportunities that 'blanket' approaches cannot.

Stream 2: Management Process Review
- Appraisal of current inventory management practices.
- Identifies the issues and behaviours which have a negative impact on inventory levels.
- Identifies the actions required for lasting results.
- Output is an action plan to address issues and behaviours.

| Stream 1 | Train | Apply | Review | Apply | Review |

| Stream 2 | Appraisal | Review Issues | Confirm & Agree | Action Plan | Implement |

Figure 9-1: The Inventory Process Optimization™ Method

2. **Utilize 7 Actions for Inventory Reduction.**
   The previous chapter discussed the 7 actions for Inventory Reduction. Now the Inventory Process Optimization™ Method turns inventory review on its head by matching the required action with the attributes of specific inventory items. This application of the 7 Actions is called the Inventory Cash Release® Process and is a key sub-process of the method.

   Rather than using a single criterion (such as obsolescence) and looking for inventory that matches the criterion, the Inventory Cash Release® Process starts by looking at individual inventory items and matching the solution to the attributes of the item. This is possible because the focus is on the high dollar values only. Therefore, we are reviewing only a small proportion of the inventory, but that review focuses on the items that will have the greatest impact on overall inventory value. In addition, this approach ensures that the solutions are specific to the item attributes. This is because the 7 Actions were developed from first principles to identify the steps for inventory reduction that can be

156

taken in the areas of supply chain, procurement, and inventory management and so we know that all possible solutions are considered.

3. **Involve a wide team and build their skills.**
The key to the ongoing application of the Inventory Cash Release® Process is training your team and working with them in the application of the process. This step has two effects. First, it engages the team in the solution so that they gain ownership of the outcomes and are better placed to champion the changes that may be necessary. Second, it builds the team's skills and knowledge in this area so that they can continue addressing inventory issues on an ongoing basis.

Stream 2: Management Process Review.

This review involves understanding your current management practices and behaviors and identifying those that have a negative impact on your inventory holdings. The steps are:

1. Appraisal of your current inventory management practices.
2. Identification and agreement with the team on the issues and behaviors that have a negative impact on inventory levels.
3. Identification and agreement with the team on the actions required for achieving lasting results.

The output of the Management Process Review is an action plan and timetable to address issues and behaviors.

Stream 1: The Individual Parts Review

The Inventory Cash Release® Process

The Inventory Cash Release® Process is a step-by-step approach that is used as part of Stream 1: The Individual Parts Review in the Inventory Process Optimization™ method. As already mentioned, the process helps companies to improve their materials management and safely, selectively, and efficiently reduce their expenditure on materials and spare parts. The process uses the Pareto Principle to focus on items with the greatest value tied up in inventory; it then applies supply chain and process review principles to what is commonly thought of in either mathematical or technical terms. This different emphasis achieves significant results because it identifies opportunities that traditional optimization does not and, in fact, cannot achieve.

The objective of all inventory reduction is to reduce inventory without disrupting the customer supply promise. That is also true with this method. That means you must not lose sight of the need to continue to meet demand. Otherwise, your inventory review will be deemed a failure and all your work will ultimately be undone. Experience shows that there are usually sufficient opportunities for genuine, risk-free reduction that you will not need to risk supply to achieve your goals (see Chapter 11: Case Studies).

Getting the Right Focus: The Pareto Principle

Vilfredo Pareto was an Italian economist who was born in 1848 and died in 1923. He is most famous for developing the 'law of the vital few,' which is also known as the Pareto Principle or the 80:20 rule. More correctly stated the Pareto Principle is:

A minority of inputs produces the majority of results.

To develop his ideas, Pareto studied the spread of wealth in Italian society at the end of the 19th century and concluded that a minority of people held the majority of the wealth. Subsequently, it was observed that the Pareto Principle held for more than just the distribution of Italian wealth. Examples of the Pareto Principal at work in modern business include:

- 80% of sales comes from 20% of product
- 80% of downtime is caused by 20% of equipment

The Pareto Principle is of great importance when reviewing any business issue, but is particularly helpful when reviewing inventory. However, using the Pareto Principle in inventory is not new. For many years, it has been advocated as an approach to use for the A, B, C analysis of inventory. The Pareto Principle has also been used to evaluate demand for inventory. Neither of these approaches helps us, however, with the task of inventory reduction.

For inventory reduction, you need to focus on the value of inventory held. Put simply, the Pareto Principle tells us that the majority of inventory value will be held in a minority of items. That is, a small percentage of inventory items will actually be accountable for the vast majority of the investment. For you this means that rather than work through your entire catalogue at a detailed level, the greatest impact in achieving an inventory reduction can be achieved by reviewing the vital few items that add most to the dollar value invested in inventory.

To determine how your inventory splits by the Pareto Principle, you need to develop a Cumulative Value Curve similar to that shown in Figure 9-2. In this figure, the horizontal axis represents the SKUs held in inventory, sorted by the dollar value held, that is the number of the SKU multiplied by the cost of the SKU. The vertical axis represents the cumulative value of these items as you move along the horizontal axis. In this example, 20% of the items account for 80% of the dollars invested in inventory.

Appropriate use of the Pareto Principle clearly demonstrates that trying to address all inventory items is not efficient, even if software is used. Maximum efficiency comes by focusing on these vital few items.

For clarity, the key to this idea is to focus on the 'high dollar value held' items. These are not the SKUs with the greatest individual value, but rather the SKUs where the total dollar value held is the greatest, that is, the cost of the SKU multiplied by the number of items held. For example, an SKU valued at $10,000 with a single item in stock requires the same investment as an SKU valued at $100 with 100 items in stock.

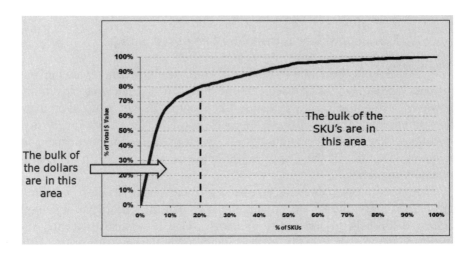

Figure 9-2: Cumulative Value Curve

Use of the Pareto Principle to identify the 'high dollar value' items that really matter is an absolute key for the Inventory Cash Release® Process and must be understood in order to gain maximum value from an inventory reduction program.

The focus on 'high dollar value held' items ensures that you concentrate on the items that tie up the bulk of the cash. It makes sense to focus on these items as the purpose of any inventory reduction program is, first and foremost, to reduce the investment of cash. In addition, using the Pareto Principle as described makes your efforts more efficient because it takes the same time and effort to review a low dollar value SKU as it does to review a high dollar value SKU.

Figures 9-3 and 9-4 show further examples of the Pareto Principle in practice.

In Figure 9-3 Company A had inventory that included 4,879 SKUs. After applying the Pareto Principle in the method described, it was recommended that they focus on just the top 250 items. These represented slightly more than half of the dollar value of their inventory but just 5% of their SKUs. In terms of their effort, using the Pareto Principle provided a reduction in workload of 95%.

In Figure 9-4 Company B had an even more dramatic result: for them, 61% of the value of their inventory was held in just 2% of the line items! (Again the focus was on was on the top 250 items by value.)

Both these organizations were able now to concentrate on the vital few SKUs and embark on an inventory reduction program that had previously seemed almost impossible.

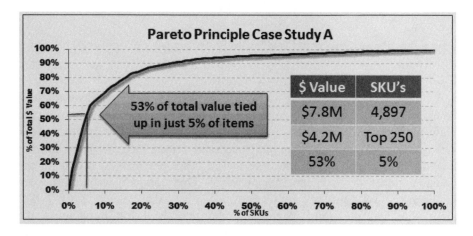

Figure 9-3: Pareto Principle Case Study A

There is no hard and fast rule about where to draw the line for the Pareto review. A review of 250 SKUs is very manageable in a short period of time. Some companies base their review on the top 500 SKUs. It is recommended that you commence by reviewing the top 250 SKUs and then progress from there.

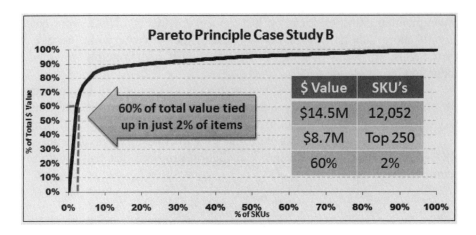

Figure 9-4: Pareto Principle Case Study B

To apply the Pareto Principle successfully, you must create a list of your inventory based on the dollar value held. Then start by reviewing the highest dollar value held SKU and work your way down the list from there. The decision on where to stop really depends on the range of items in your inventory, their value, and the time that you are prepared to invest for the potential return on each incremental item. One company decided to keep reviewing SKUs until they got down to a 'dollar value held' of $1,000. For them, this was the point where they felt that the effort outweighed the benefit. Their review to this level took several months and cost over $100,000 in labor and overtime payments but realized more than $3,000,000 in inventory reduction — a payback on expense of 30:1!

Having identified the key SKUs to work on, you will next apply the 7 Actions for Inventory Reduction. To make the selection of which action to apply a little easier, it helps if, before you begin, you have categorized or segmented the items that are in the high value segment (the Pareto Inventory).

Categorization also helps to ensure that you don't focus solely on the fast moving items. As explained previously, fast movers become a focus because they seem to represent the shortest path to inventory reduction. The logic is that if you stop ordering fast movers, your stock will fall quickly. Although the mechanics of this approach are true, it is risky as it may lead to stockouts. It also ignores the fact that the high value segment will also contain low turnover, slow moving, possibly obsolete, and potentially phantom stock entries. Depending upon the approach taken, these items can also quickly reduce the actual and book value of investment in inventory. Note that the categorization here differs from that required for developing stocking policies. The categories required here are based, typically, on inventory movements and help identify which of the 7 actions to apply and the likely time frame for benefit realization. Table 9-1 defines some categories that might apply to MRO inventory.

The actual categories for your business and inventory types might be different than these, as might the time frames used to define high turnover and low turnover items. In the section *When to Apply Which Actions* (see below) the definitions in Table 9-1 are used to demonstrate how categories help determine which of the 7 Actions to apply. You should check that these categories apply to your situation and interpret the data accordingly.

Category	Definition	Potential Sub Element	
High Turnover	+3 movements per annum	Excess — in excess of 2 years stock held.	Surplus — current stock is above control level
Low Turnover	At least one movement in 4 years (but less than 3 per annum)		
Slow Moving	No movement in 4 years		
Obsolete	No longer required		
Phantom	Part and/or location unknown		

Table 9-1: Categories for Spares Inventory

When to Apply Which Action

There is one more insight that will help you to minimize your efforts in the application of the 7 Actions and it is this: Because of the nature of the actions and the nature of inventory, it is relatively easy to predict which actions would most likely be applied to which categories of inventory.

Using the categories from Table 9-1, the 7 Actions and the type of inventory to which each might be applied is summarized in Figure 9-5. Down the left hand side are listed the various types of inventory as per the previous definitions. Across the top are the actions that can be applied. Note that it is entirely possible that more than one action could be taken for any particular inventory item. The X marks indicate that the particular action is most definitely applicable to the inventory type and that the action is likely to have a timely impact. The shaded squares indicate that the action may be applicable and/or may take some time to take effect. If you have developed a list of inventory types that is specific to your business, you should now prepare your own reference table to help guide when to use which action.

Once you have your inventory data and have identified your inventory categories, the chart can be read in two ways. First, you can read across from the inventory type to see which actions may be applicable. For example, High/Low Turnover inventory can be influenced by almost all of the actions; you can expect these actions to have an impact in a reasonable time frame (depending on how quickly the items move). However, you can see that Action [#]2 does not impact inventory unless it is overstocked.

Similarly, for Slow Moving stock you are less likely to do a consignment deal; Actions [#]4 and [#]5 will take some time to have an impact. The best course of action for Slow Moving stock will be to Eliminate Duplication (Action [#]3) or More Closely Match Delivery With Usage (Action [#]6).

The other way to read the chart is down from the action type. This approach helps when you are assigning actions for review by members of your team. For example, Action [#]1 is likely to need the direct input from your purchasing/procurement team members. Therefore, they know that they should pay attention to what can be done for the items classified High/Low Turnover. Similarly, appropriate individuals can now be assigned to Eliminate Duplication and so on. Using the categories in this way helps to further break down the workload for the individuals. Instead of reviewing (say) 5,000 items, the focus moved to 250 items. By using the chart to focus the specific action of individual team members, they may need to focus only on 30 or 40 items.

Finally, the inventory type of Phantom is included despite the fact that there is little that you can do about Phantom inventory. Phantom items are where the item has already been used or sold but has not been signed out of your inventory. The only course of action that you can take is to expense the item. (In the table, we have chosen to include expensing the item under Action [#]2, Sell Excess and Obsolete Stock.) Expensing the items might create some short-term budgetary pain, but should be used as an opportunity to highlight systematic problems in your inventory management.

Type of Inventory	Action						
	1	2	3	4	5	6	7
High/Low Turnover	X		X	X	X	X	X
Slow Moving	▨		X	▨		X	
Obsolete		X					
Overstocked		X	X	▨		X	
Phantom		X					

X = Definitely review ▨ = Possibly review

Figure 9-5: Summary of When to Apply Which Action

The Step-By-Step Process

The Inventory Cash Release® Process is a step-by-step process that acts as a decision making filter. When there are a large number of inventory items, it seems that an inventory review requires a large number of decisions. The use of the Pareto Principle reduces this through considered selection of the items to review and the decisions required are then further refined through the application of one or more of the 7 Actions. This process produces a smaller number of focused decisions that will make a real difference to your inventory holdings. This is shown schematically in Figure 9-6.

One of the major mistakes that people make when applying the Inventory Cash Release® Process is that they 'cherry pick' the actions that appeal to them. You absolutely must not do this. Cherry picking shortcuts the process but also means that you are pre-deciding which actions are applicable. The risk is that you will miss out on other real opportunities that you may not have thought of.

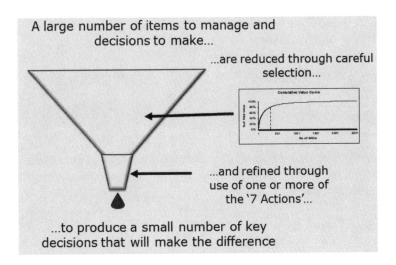

A large number of items to manage and decisions to make...

...are reduced through careful selection...

...and refined through use of one or more of the '7 Actions'...

...to produce a small number of key decisions that will make the difference

Figure 9-6: The Process Acts as a Decision Making Filter

To explain the Inventory Cash Release® Process, the following description refers to Figure 9-7. Start at the position marked as A.

A: Having compiled your list of SKUs based on the 'dollar valued held' and drawn your Pareto curve, you start the review by focusing on the SKU at the top of your list. This is the SKU in which you have the most funds invested.

You now determine the category of the SKU based on the category types that you developed previously. (Later you will see how this categorization can help you to determine which of the 7 Actions may be most applicable.)

B: Now consider if you can apply Action [#1] to this SKU. At this stage you don't need to definitively know the answer. Just decide if you think it is a possibility worth pursuing. If the answer is 'yes,' move on to C; if the answer if 'no,' move on to E.

C: Now you should identify your plan to review the application of the action or the implementation of the action if you have definite ideas. Earlier in this book, there were checklists for each action that you can use as the basis of your plans.

D: Now consider if further actions can be applied to this SKU. If the answer is 'yes,' move on to E; if the answer is 'no,' move onto G.

E: Now repeat the logic of steps B, C, and D for each of the 7 Actions. This will systematically work you through each of the 7 Actions and their applicability to this specific SKU. Remember that at this stage you are still considering just the one SKU.

F: After you have considered each of the 7 Actions, you might realize that no action can be taken — this is OK. One of the misconceptions about inventory reviews is that you must reduce your holdings of each SKU by a little bit. The reality is that most reviews identify big opportunities with a small number of items.

G: By the time you reach step G you will have considered all of the opportunities for the SKU in question. At this point you should document your decisions and actions determined thus far.

H: Now turn your attention to the next SKU on the list and consider if the value of the SKU is sufficient for further effort to be spent on scrutiny. In the beginning of your review, the SKUs will have sufficient funds tied up to warrant further review. However, you will reach a point where you decide that the time and expense spent in review is greater than the benefit that could be realized.

If the decision is that the value is sufficient for further scrutiny, then go back to step B and restart the process for the next SKU.

If the decision is that the value is not sufficient for further scrutiny, then move onto step I.

I: By this step you will have reviewed all of the SKUs that matter and the application of each of the 7 Actions. Now you should compile your overall plan for implementation, including time lines, responsibilities and reporting on progress. Now you can get on with implementation.

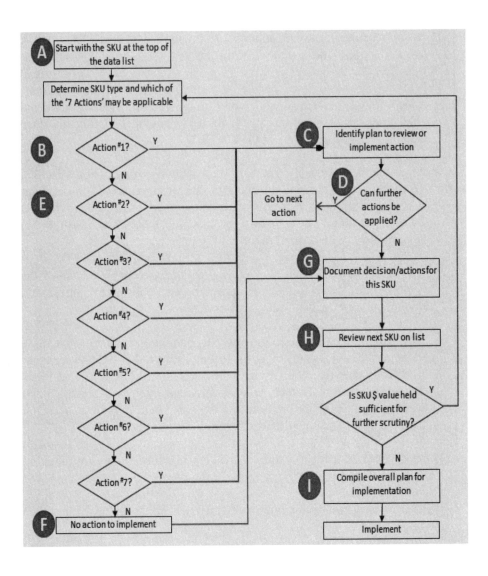

Figure 9-7: The Step-by-Step ICR® Process

Example: Application of the ICR® Process

The Inventory Cash Release® Process provides a systematic approach to identify the vital few items for attention, and then address those items with one (or more) of the 7 Actions for Inventory Reduction. Applying the Inventory Cash Release® Process requires the collation of appropriate data and some decision making. Typically this involves developing a spreadsheet to lay out the relevant data and record the action to be taken.

To complete this chapter, we are going to go step-by-step through an example. Included here are several pages that show each step in the process. These are labeled as Figures 9-8, 9-9, 9-10, 9-11, and 9-12. The example here is for engineering spares, but the spreadsheet could be equally applied to other types of inventory

Here goes!

Once you have been assigned the job of reviewing your inventory, the first step that everyone takes is to create a spreadsheet. If you are like most people who approach this task without applying the Inventory Cash Release® Process, you may well have created a spreadsheet like the one in Figure 9-8.

Take a minute or two to review Figure 9-8. It has several columns: Item No., Stock Code, Description, Quantity on Hand, and Item Value. For the purpose of this example, Figure 9-8 is sorted by both Stock Code and Item Value.

In most cases, people sort their inventory by some classification (such as Stock Code or Item Value). Their logic is that the Stock Code sort will enable them to hand over the list to the key users responsible for that inventory and have them do a review. In this case, the users might be the engineers responsible for specific spares inventory. This approach leads people down the path of reviewing the high value items first; as we now know, this is not appropriate for an effective and efficient process.

Sorting by Stock Code and assigning a review based on this sort, potentially results in the review of low-value items or stock where there is only a small inventory investment. As it takes as much time to review items with a low inventory investment as it does items with a high inventory investment, you should focus your time and limited resources where the higher investment exists.

However, sorting by item value is not the answer. It can result in looking at the cost of each item rather than the total investment. Again, the use of time and resources in this way will be less efficient.

Now turn to Figure 9-9. In this figure the table has been sorted by 'Bin Value'. To calculate this value, multiply the Quantity on Hand by the Item Value. In Figure 9-9, the table has been sorted by Bin Value from highest to lowest. You can see that the first two items have significantly lower individual item values than the third item, but involve a larger number of components and subsequently a higher total investment. We know that we are now focusing on the items where our cash is tied up!

Although these first two items have significant item value, notice that the fourth item on the list has a relatively low item value. Again, because there are a large number of items, this item represents a significant total dollar held. Accordingly, it should be near the top of our considerations.

Now turn to Figure 9-10.

In this figure the data is sorted by total dollar value held. You can generate a Cumulative Value Curve like the one shown in Figure 9-10. To do this, you need to include another column (not shown here) and use that column to calculate the cumulative value of the inventory as you go down the list.

For this column, row 1 will be $48,000
Row 2 will be $94,000 ($48,000 + $46,000)
Row 3 will be $114,000 ($94,000 + $20,000), and so on.

Figure 9-10 demonstrates clearly that the majority of inventory value is held in a minority of items (the Pareto Inventory). This is where to focus your attention. You can use this curve to help convert any doubters in your business.

Now turn to Figure 9-11.

We have now added two more columns to the previous spreadsheet: Category and Action to Apply. The first step in working through the list is to identify the category in which each item fits. As per the discussion above, identifying the category helps guide us in knowing which action to apply. It will also help understand the potential time frame for implementation. Realistically, a slow-moving item may have its reorder parameters changed, but because it is slow moving it may take months or years for the change to have a real impact.

You now need to consider which action to apply to each item. Remember, it is possible that you can apply more than one action to an item. For this example, however, we have applied only one action to each item.

Working from the top down, each item is considered in turn and the appropriate action determined. (Don't worry if the item descriptions are unfamiliar to you. It is the process that matters.)

For example:

Item 1 is a Delivery Valve where the turnover is low. But in this case, the full set of 24 items is considered as critical. It has been decided to take no action on item 1. The same goes for Items 2 and 3.

Item 4 is a Thyristor device. These are a high turnover item and so may be suitable for consignment stocking from the vendor. This is noted in the table.

Item 5 is an item that is thought to be OK in terms of inventory level.

Item 6 is a Pump that is categorized as slow moving. We notice that there are 3 of them in stock. Holding 3 may be too many and so this item is noted as likely to be sold as overstocked.

Moving down the list, Item 9 is a coupling where the usage is quite predictable, so perhaps they don't need so many on hand. They decided to work to matching delivery more closely with usage.

And so on.

I trust that you can see how the combination of classifying the item and deciding which action to apply can help you then to assign to the appropriate team members the tasks of reviewing the inventory and negotiating with suppliers.

Now turn to Figure 9-12.

Now we have added three more columns: Comment, Projected Value, and Inventory Reduction.

The Comment column allows you to include a comment or define the action required. In this example, the column is included as a summary. It should not be seen as a replacement for properly documenting the tasks, the responsibility, and the time frame.

Read down this column and you will notice that none of the actions are extreme. Look through the data and comments; consider the circumstance of each item. You will notice that in each case the action appears to be quite reasonable. This is because the actions are taken in context of any changed circumstance and are designed to reduce inventory without impacting risk.

The Projected Value column is the predicted value of the inventory once the tasks assigned for reducing the inventory are complete.

The Inventory Reduction column is the difference between the Bin Value and the Projected Value; it tells us by how much the inventory is projected to reduce.

Obviously this spreadsheet is simplified. For our purpose, this spreadsheet is designed to show how the process is applied. When working through a real spreadsheet, though, any number of other columns could be added. They might include:

- Other data such as the maximum and minimum holdings to highlight overstock and high safety stocks
- More detail on the action to be taken, by whom, and when
- The potential timing of the impact of the action
- Whether the action will affect the book value of the inventory

The spreadsheet shown is only meant to be a summary. It should not replace a more complete tracking of actions or even a time line that shows the period over which the inventory reduction will occur.

Finally, look at Figure 9-13.

This figure is a screen shot of software that has been developed specifically to support the application of the Inventory Cash Release® Process. This software helps with both data and project management; it is much simpler to use than a series of separate spreadsheets. Details can be found at the website www.PhillipSlater.com, or by contacting the author directly.

Item No.	StockCode	Description	Qty on Hand	Item Value	Bin Value
1	AA1234	Intercooler	1	$ 20,000	$ 20,000
2	AA1235	Gearbox	1	$ 12,000	$ 12,000
3	AA1236	Motor	2	$ 9,000	$ 18,000
4	AA1237	Motor	1	$ 8,500	$ 8,500
5	AA1238	Gearbox	2	$ 6,500	$ 13,000
6	AA1239	Clutch	1	$ 5,600	$ 5,600
	240	Pump	3	$ 4,500	$ 13,500
	241	Spindle	3	$ 2,70	
	242	Pinion	3	$ 2,50	
	243	Delivery Valve	24	$ 2,00	
11	AA1244	Suction Valve	24	$ 1,95	
12	AA1245	Flange	1	$ 1,80	
13	AA1246	Load Cell	3	$ 1,700	$ 5,100
14	AA1247	Hydraulic Cylinder	5	$ 1,600	$ 8,000
15	AA1248	Coupling	8	$ 1,500	$ 12,000
16	AA1249	Casing	1	$ 1,500	$ 1,500
17	AA1250	Impeller	1	$ 1,500	$ 1,500
18	AA1251	Chain Assmbly	1	$ 1,250	$ 1,250
19	AA1252	Support Pipe	1	$ 1,200	$ 1,200
20	AA1253	Cylinder Hydraulic	3	$ 750	$ 2,250
21	AA1254	Piston and Rod	4	$ 575	$ 2,300
22	AA1255	Thyrister Equip	56	$ 350	$ 19,600
23	AA1256	Hose Assembly	6	$ 300	$ 1,800
24	AA1257	Sensor	35	$ 275	$ 9,625
25	AA1258	Hydraulic Fitting	15	$ 250	$ 3,750
26	AA1259	Bearing, tapered	10	$ 250	$ 2,500
27	AA1260	Cartridge Filter	6	$ 250	$ 1,500
28	AA1261	Cylinder Bushing	16	$ 80	$ 1,280
29	AA1262	Circuit breaker	25	$ 75	$ 1,875

Sorted by stock code

Or sorted by item value

Figure 9-8: Typical Inventory Review Spreadsheet

173

Item No.	Stock Code	Description	Qty on Hand	Item Value	Bin Value
1	AA1243	Delivery Valve	24	$ 2,000	$ 48,000
2	AA1244	Suction Valve	24	$ 1,950	$ 46,800
3	AA1234	Intercooler	1	$ 20,000	$ 20,000
4	AA1255	Thyrister Equip	56	$ 350	$ 19,600
5	AA1236	Motor	2	$ 9,000	$ 18,000
6	AA1240	Pump	3	$ 4,500	$ 13,500
7	AA1238	Gearbox	2		13,000
8	AA1235	Gearbox	1	Sorted by	12,000
9	AA1248	Coupling	8	bin value	12,000
10	AA1257	Sensor	35		9,625
11	AA1237	Motor	1	$ 8,500	$ 8,500
12	AA1241	Spindle	3	$ 2,700	$ 8,100
13	AA1247	Hydraulic Cylinder	5	$ 1,600	$ 8,000
14	AA1242	Pinion	3	$ 2,500	$ 7,500
15	AA1239	Clutch	1	$ 5,600	$ 5,600
16	AA1246	Load Cell	3	$ 1,700	$ 5,100
17	AA1258	Hydraulic Fitting	15	$ 250	$ 3,750
18	AA1259	Bearing, tapered	10	$ 250	$ 2,500
19	AA1254	Piston and Rod	4	$ 575	$ 2,300
20	AA1253	Cylinder Hydraulic	3	$ 750	$ 2,250
21	AA1262	Circuit breaker	25	$ 75	$ 1,875
22	AA1245	Flange	1	$ 1,800	$ 1,800
23	AA1256	Hose Assembly	6	$ 300	$ 1,800
24	AA1249	Casing	1	$ 1,500	$ 1,500
25	AA1250	Impeller	1	$ 1,500	$ 1,500
26	AA1260	Cartridge Filter	6	$ 250	$ 1,500
27	AA1261	Cylinder Bushing	16	$ 80	$ 1,280
29	AA1251	Chain Assmbly	1	$ 1,250	$ 1,250
28	AA1263	Grinding Wheels	25	$ 50	$ 1,250

Figure 9-9: An ICR® List Sorted by Bin Value

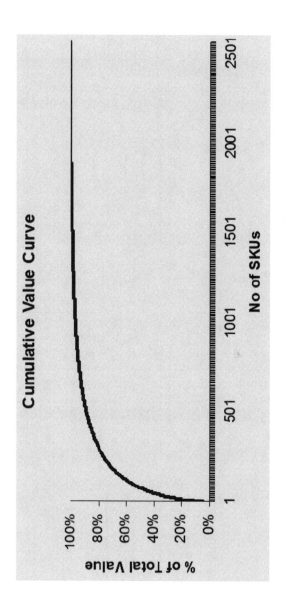

Figure 9-10: Example Cumulative Value Curve

Item No.	Stock Code	Description	Qty on Hand	Item Value	Bin Value	Category	Action to Apply	Comment
1	AA1245	Delivery Valve	24	$ 2,000	$ 48,000	Low Turnover	Nil	Critical spare with no alternate
2	AA1244	Suction Valve	24	$ 1,950	$ 46,800	Low Turnover	Nil	Critical spare with no alternate
3	AA1254	Intercooler	1	$ 20,000	$ 20,000	Slow Moving	Nil	Critical spare with no alternate
4	AA1255	Thyristor Equip	56	$ 350	$ 19,800	High Turnover	1	Put on consignment
5	AA1226	Motor	2	$ 9,000	$ 18,000	Slow Moving	Nil	Repairable item
6	AA1240	Pump	3	$ 4,500	$ 13,500	Slow Moving	2	Sell one pump
7	AA1238	Gearbox	2	$ 6,500	$ 13,000	Slow Moving	Nil	Repairable item
8	AA1235	Gearbox	1	$ 12,000	$ 12,000	Slow Moving	Nil	Repairable item
9	AA1236	Coupling	8	$ 1,500	$ 12,000	Slow Moving	6	Coordinate delivery with ...
10	AA1257	Sensor	35	$ 275	$ 9,625	High Turnover	1	Put on consignment
11	AA1237	Motor	1	$ 8,500	$ 8,500	Low Turnover	Nil	Critical spare with no alternate
12	AA1241	Spindle	1	$ 8,500	$ 8,500	Obsolete	2	Remove item
13	AA1247	Hydraulic	4	$ 1,800	$ 8,000	Low Turnover	4	Reduce maximum 5-3 due to supplier changes
14	AA1242	Pinion	3	$ 2,500	$ 7,500	Low Turnover	5	Reduce reorder by 1
15	AA1239	Clutch	1	$ 5,600	$ 5,600	Obsolete	5	Remove item
16	AA1246	Load Cell	3	$ 1,700	$ 5,100	Low Turnover	5	Reduce reorder by 1
17	AA1258	Hydraulic Fitting	15	$ 250	$ 3,750	High Turnover	4	Reduce maximum from 5 due to supply chain changes
18	AA1259	Bearing Tapered	10	$ 250	$ 2,500	High Turnover	3	Duplicated with item AA3458 - remove this item
19	AA1254	Piston and Rod	4	$ 575	$ 2,300	Low Turnover	4	Reduce maximum of 2 due to supplier changes
20	AA1253	Cylinder Hydraulic	3	$ 730	$ 2,250	Obsolete	2	Remove item
21	AA1252	Circuit breaker	25	$ 75	$ 1,875	High Turnover	1	Put on consignment
22	AA1245	Flange	1	$ 1,800	$ 1,800	High Turnover	Nil	Critical spare with no alternate
23	AA1256	Hose Assembly	6	$ 300	$ 1,800	High Turnover	1	Put on consignment
24	AA1249	Casing	1	$ 1,500	$ 1,500	Slow Moving	Nil	Critical spare with no alternate
25	AA1260	Impeller	1	$ 1,500	$ 1,500	Slow Moving	8	Critical spare with no alternate
26	AA1260	Cartridge Filter	8	$ 255	$ 1,500	Slow Moving	8	Coordinate usage with PM
27	AA1281	Cylinder Bushing	16	$ 80	$ 1,280	Low Turnover	5	Reduce reorder so max is 8
28	AA1251	Chain Assembly	1	$ 1,250	$ 1,250	Low Turnover	Nil	Critical spare with no alternate
29	AA1255	Grinding Wheels	25	$ 50	$ 1,250	High Turnover	7	Purchase from new supplier at half price

Category

First Pass
Action to
Apply

Figure 9-11: An Expanded Spreadsheet

Item No.	Stock Code	Description	Qty on Hand	Item Value	Ext Value	Category	Action to Apply	Comment	Projected Value	Inventory Reduction
1	AA-243	Delivery Valve	24	$ 2,000	$ 48,000	Low Turnover	Nil	Critical spare with no alternate	$ 48,000	$ -
2	AA-244	Suction Valve	24	$ 1,950	$ 46,800	Low Turnover	Nil	Critical spare with no alternate	$ 46,800	$ -
3	AA-234	Intercooler	1	$ 20,000	$ 20,000	Slow Moving	Nil	Critical spare with no alternate	$ 20,000	$ -
4	AA-256	Thyrister Equip	56	$ 350	$ 19,600	High Turnover	1	Put on consignment	$ -	$ (19,600)
5	AA-236	Motor	2	$ 9,000	$ 18,000	Slow Moving	Nil	Repairable item	$ 18,000	$ -
6	AA-240	Pump	3	$ 4,500	$ 13,500	Slow Moving	2	Sell one pump	$ 9,000	$ (4,500)
7	AA-238	Gearbox	2	$ 6,500	$ 13,000	Slow Moving	Nil	Repairable item	$ 13,000	$ -
8	AA-235	Gearbox	1	$ 12,000	$ 12,000	Slow Moving	Nil	Repairable item	$ 12,000	$ -
9	AA-248	Coupling	8	$ 1,500	$ 12,000	Low Turnover	8	Coordinate delivery with supplier	$ 3,000	$ (9,000)
10	AA-257	Sensor	35	$ 275	$ 9,625	High Turnover	1	Put on consignment	$ -	$ (9,625)
11	AA-237	Motor	1	$ 8,500	$ 8,500	Low Turnover	Nil	Critical spare with no alternate	$ 8,500	$ -
12	AA-241	Spindle	3	$ 2,700	$ 8,100	Obsolete	2	Remove item	$ -	$ (8,100)
13	AA-247	Hydraulic Cylinder	5	$ 1,600	$ 8,000	Low Turnover	4	Reduce maximum to 3 due to supplier changes	$ 4,800	$ (3,200)
14	AA-242	Pinion	3	$ 2,500	$ 7,500	Low Turnover	5	Reduce reorder by 1	$ 5,000	$ (2,500)
15	AA-239	Clutch	1	$ 5,600	$ 5,600	Obsolete	2	Remove item	$ -	$ (5,600)
16	AA-246	Load Cell	3	$ 1,700	$ 5,100	Low Turnover	5	Reduce reorder by 1	$ 3,400	$ (1,700)
17	AA-258	Hydraulic Fitting	15	$ 250	$ 3,750	High Turnover	4	Reduce maximum to 5 due to supply chain changes	$ 1,250	$ (2,500)
18	AA-254	Bearing Tapered	10	$ 230	$ 2,300	High Turnover	3	Duplicated with item AA-3466 - remove this item	$ -	$ (1,150)

Figure 9-12: The Completed Spreadsheet

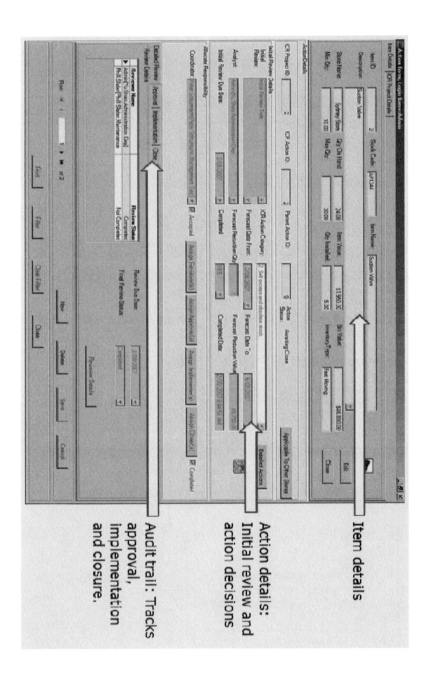

Figure 9-13: Software Makes It Easier to Track the Process

Reviewing the Long Tail

The 'long tail' refers to the remaining inventory items that lay outside of your Pareto items (see Figure 9-14). Working through these items is problematic because, by definition, they are of relatively little value. However, this does not mean that they should be ignored. Rather than jump in and review these items on a one-by-one basis, there needs to be a further filter applied to ensure that only those that present real opportunity are reviewed. This filter is to identify those items that we know add no value; primarily these will be the items that are overstocked. It is also likely that there will be items that are obsolete.

The process by which items become overstocked is exactly the same as that discussed previously—returns, aggressive re-ordering, and change to the dynamics of supply and demand. Once the items that are overstocked are identified, the approach to managing the quantity down can be determined. The approach could include allowing natural attrition, preferential use, or sales of excess stock. As always, prioritize the items to work on based on total dollar value.

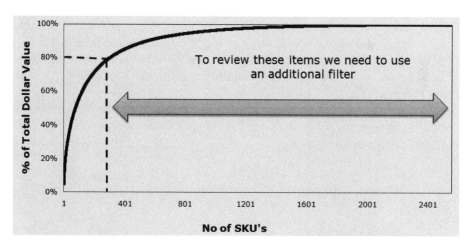

Figure 9-14: Identify the Long Tail of Items that Add No Value

There are two ways that overstocked items can be identified. The first is to use your inventory management system to generate a 'stock over maximum' report. This report should identify all those items where the stock currently held is in excess of the predetermined or natural maximum.

A predetermined maximum exists if your system uses a max-min approach for managing inventory. A natural maximum exists if you use an ROP/ROQ approach. The natural maximum is your ROP + ROQ. In theory, you should never hold more than this number.

The second way to identify overstocked items is to generate a 'dead stock' report. This report identifies where the SKU in question has never reached zero in stock. Figure 9-15 shows both stock over max and dead stock situations.

Potentially obsolete (or probsolete) stock is identified by reviewing those items with no turnover for some period. This period for review depends upon the quantity involved and the effectiveness of your normal review on obsolete stock. If the normal demand on your stock is generally high, you may set the review target at, say, no movement for 1 year. If you routinely have items that don't move for long periods, as in an engineering situation, you may set this target at 4 or more years. The obsolete items are managed as described previously.

When you generate either (or both) reports, you then need to apply almost exactly the same process as for the Pareto inventory. The difference is that here you have more choice. You can either sort your list by bin value (as for the Pareto inventory) or you can sort it by the size of the opportunity, that is, the amount of overstock or dead stock. Obviously obsolete items should be sorted by bin value (by definition the bin value will be the size of the opportunity).

Now that you have a refined list that is sorted by value, you can once again begin working from the high value/opportunity items and one-by-one decide your actions.

Example of Long Tail Inventory Reduction

Stock Over Max.

Out of an inventory comprising nearly 8,000 items, this company targeted the top 450 items in their inventory as their Pareto review. They then ran a report that identified a further 1,200 items which were overstocked. The combined value of the overstocked quantity was nearly $3M!

While reviewing 1,200 items is a big task, knowing that there is a prize of $3M for successful completion helps prioritize which items to review and the appropriate resources to apply.

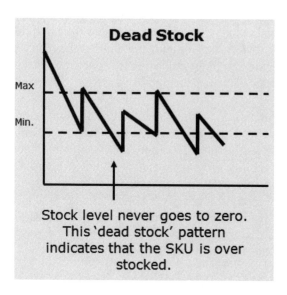

Dead Stock

Stock level never goes to zero. This 'dead stock' pattern indicates that the SKU is over stocked.

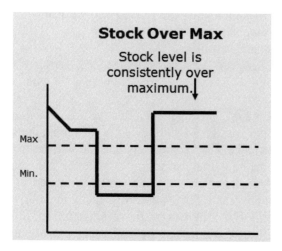

Stock Over Max

Stock level is consistently over maximum.

Figure 9-15: Identifying Overstocked Items

Stream 2: The Management Process Review

The previous sections have discussed the Individual Parts Review associated with Inventory Process Optimization™. However, having established what changes are required for individual parts, it is important to now review, understand, and address the management processes that will have an ongoing impact on your materials and inventory management results. This is achieved in Stream 2: the Management Process Review.

The importance of this stream lays in two concepts discussed previously: Systems Thinking, discussed in Chapter 4; and Home Truth [#]1 (Inventory Does Not Exist in Isolation), discussed in Chapter 6. At stake here is the sustainability of your inventory review. By sustainability, I mean the ability of your results to be sustained over a significant period of time. Ideally, you want to make a complete culture change so that the problems do not return at all; this is where you start to make those kinds of changes stick.

The key steps of the Management Process Review are:

1. Appraisal of your current inventory management practices
2. Review of the issues identified
3. Confirmation and agreement with your team
4. Development of your action plan
5. Implementation

The interaction of these steps in the total process is shown in Figure 9-1.

One of the major challenges with this type of review is that it relies on the experience and ability of the reviewer in recognizing problems and issues. This experience is not something that you will get from just reading a book; this is something that comes from practice and application. It is suggested, therefore, that you study the issues addressed in the first half of this book and then practice by observing them in your own workplace. Because this part of the Inventory Process Optimization Method is less process driven, further discussion on undertaking a Management Process Review is included in Chapter 10, Execution: Taking Action to Achieve Results.

Chapter 10

Execution: Taking Action to Achieve Results

Here is perhaps the most important statement in this book: Everything that you have reviewed and learned so far will amount to nothing if you do not take steps to implement changes in your materials and inventory management. While improving knowledge is important, knowledge without action adds little or no value. Execution requires more than just rolling out the steps detailed in this book. It involves managing the process of the roll out (that is, scheduling, training, follow up, documentation, and so forth) and managing the change associated with the roll out (attitudes, ideas, policy, procedures, metrics, and acceptance). Unless the process and change are managed correctly, you will be unlikely to achieve true inventory optimization.

This chapter deals primarily with the process of execution and is not intended to provide insight into managing the change — that is a whole book topic on its own and there are plenty of good books that deal with that. This chapter will, however, take you through a known and proven process and provide some guidelines for implementation.

Before embarking on an implementation program, there are three questions that you need to know the answers to:

1. Can you measure your inventory investment?
   Some companies expense items when they are purchased and, therefore, do not have a separate allocation that identifies inventory. To apply the Inventory Cash Release® Process successfully, your

company must have identified its investment in inventory as a separate allocation (i.e., it must be capitalized or at least traceable).

2. Are your systems capable of generating the required data?
   As you know by now, collecting the right data is important for completing your Pareto review. Therefore, not only will you need to be able to collect the inventory data, but also you will need to have a means of readily sorting the inventory data. Most computer-based systems will enable the user to sort the inventory data in the required fashion; if you can measure your inventory investment, you can probably also sort it. Best to double check before you commence.

   If you cannot collect the right data, then your ability to apply the process will be limited. You will still be able to apply the Stock Decision Checklist in this chapter and make better decisions about the inventory you hold. However, your efficiency and effectiveness will be reduced as these fundamentally come from the initial data collection.

3. Do you have the right level of support for an inventory reduction program?
   That is, does the program have support at the right levels to ensure that the participants prioritize their involvement? If you do not have support, then you should work on internally selling the idea of a program before trying to commence one.

   One way to generate management interest in applying the process is to understand just how much cash your company could generate through an inventory reduction program. To assist you, a calculation template is included on page 47. This template will help you to identify the cash potential from implementing an inventory reduction program in your company. By completing this template, you will understand just how much your company could gain through application of this process. This information will help you to sell the idea of the program to senior management.

If you can answer 'yes' to all three questions, then you are in a position to undertake an inventory reduction program. If your answer is 'no' to any question, then you need to do some further work before commencing the process.

Finally, before commencing with the Inventory Process Optimization™ Method, it is important that you and your management are prepared to commit to making this process work. The Inventory Cash Release® Process is proven and will deliver a lasting result if the actions are followed. However, along the way, there may be some difficult decisions to be made and some bitter pills to swallow (such as writing off unnecessary inventory). The Management Process Review requires honest and impartial evaluation of your policies, procedures and behaviors. Commitment and follow through are required if you are to achieve the greatest benefit.

Selecting Your Inventory Review Champion

Having decided that you are in a position to commence the program, the next step is to identify someone not only to run the program but also to champion the program — your Inventory Review Champion. This person takes responsibility for ensuring that the Inventory Process Optimization™ Method is implemented properly and followed up rigorously. Choosing a champion is perhaps the most important decision that you will make when implementing this or any other program. Choose the wrong champion and your program will underachieve or falter. Choose the right champion and your program will have greater chance of success and delivery of a long-term result.

There are three attributes that champions must have if they are to succeed:

1. Coordination
   Champions need to be able to coordinate and manage a group of people. This means being able to negotiate attendance at workshops and meetings, schedule the workshop or meeting, and make sure that everything in their control happens as it should. The ability to coordinate is often assumed to be a 'given'; after all, how hard can it

be? But it is actually not a universal skill and the ability of your champion should be carefully considered.

2. Passion
   Champions need to have passion for delivering sustainable materials and spares inventory management improvements. If they think that the program is not worthwhile or that they can gain kudos with a short-term result, then the rest of the participants will pick up on this attitude and your program will under-deliver.

3. Influence
   Champions need to be able to influence other people in the team to do the work required and achieve the goals. Influence involves the way they interact with people and understand the issues they face in day-to-day activity. Influence also comes from respect — the respect of the team for the champion and their ability to interpret situations and deliver the program.

The ideal champion has the right mix of each of these attributes, as shown diagrammatically in Figure 10-1.

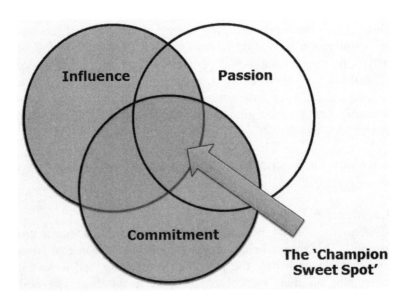

Figure 10-1: The Inventory Review Champion 'Sweet Spot'

Consider the consequences of champions not having these attributes:

- Champions who can coordinate but have no passion or influence will set meetings, but not care if people don't attend.

- Champions with passion but poor coordination or influence will make plenty of noise, will have little or no respect, and will be unlikely to get any momentum in the program. They usually default to doing all the work themselves.

- Champions with influence but no passion for the project might lead your team down the wrong path!

Selecting your champion is important. The champion is not just someone with the time to do the job. The champion is also someone who sees the value in the program, wants the team to improve, has the influence with the team, and can make it all happen!

The Role of Your Inventory Review Champion

It is important to realize that the champion's role is not to complete all the analytical work and implement the changes. Instead, the champion's role is to facilitate the development of your team and the application of the Inventory Process Optimization™ Method.

Therefore, the role is required to:

- Manage the process.
- Set the timetable and organize any training and workshops.
- Coordinate group discussions.
- Document the consensus reached by the team.
- Support the team and follow up on agreed actions.
- Identify roadblocks and assist the team in overcoming them.
- Identify existing principles, policies, measures, and reporting.

- Work with the team to develop new principles, policy, measures, and reporting.
- Report on progress/issues to the senior management that is sponsoring the program.

Champions do not need to be a professional consultant, but they must have an understanding of inventory management and the issues discussed in this book.

Once you have selected your champion, you will need to ensure that the right people are involved in your program.

Who to Involve in an Inventory Review Program

To be successful your program will require a mix of skills and input. One of the key challenges with materials and spares inventory management is that it is one of the truly cross-functional activities that a company undertakes; it involves a mix of technical, financial, purchasing, and stores personnel. This means that inventory management often falls between the cracks of technical management and working capital management. As a result, there is often a misalignment between those that decide what is held in inventory and those who know how much it really costs or how much should be invested. Your program will need to overcome these issues by including:

- At least one representative for each location participating in the program. These people will need to be able to complete the step-by-step Inventory Cash Release® Process and identify opportunities.

- At least one decision maker who can say yes or no to decisions to change inventory levels. This may be the same person as above or it may not. This person may also be called upon to make decisions on new policy, measures, and reporting.

- Any key support people who have an impact on the ability to actually implement decisions. For example, these could be IT or procurement personnel.

- A representative of your financial management team (you may call them Finance or Accounts depending on which country you are from). This person will need to be able to access financial data on your inventory as well as your depreciation and write down policies.

- A representative of the users of your inventory. The users will need to know that your review has been thorough so you may as well invite them along for the journey. For spare parts, this representative may be an engineer responsible for production.

- Your Inventory Review Champion.

The Three Stages of an Inventory Process Optimization™ Program

Once you decide to implement an Inventory Process Optimization™ program the obvious goal is to reduce your inventory investment. However, achieving this goal requires you to consider the process and changes that need to be put in place for the long-term outcome to be different from today's outcome. To achieve a result that can be sustained over the long term, your inventory review program needs to have the following implementation goals. These are:

1. Educate your team on inventory review techniques.
2. Kick start the application of these techniques so that your company starts to gain the benefits as soon as possible.
3. Establish the principles, policies, measures, and reporting to make your inventory reduction sustainable.

To achieve these implementation goals, divide your program into three stages: Preparation, Stream 1, and Stream 2 (as per the process discussed in the previous chapter). The overall process and the sequencing of activities for an implementation program are shown in Figure 10-2 and the implementation of these stages is discussed separately on the following pages.

In the author's experience, the suggested program should be conducted over a 14-week period. A timeframe of 14 weeks will allow you to achieve an appropriate balance of time between training/workshops, getting the required work done, and regular contact to maintain focus. If the time is shortened, there will be insufficient opportunity for participants to complete their tasks and do their regular day jobs. If the time is lengthened, the program typically loses focus and elements may need to be repeated.

At the conclusion of the program stages laid out in Figure 10-2, you may not have achieved all of your inventory goals but you will have started the process of change; you will have in place all the elements necessary to continue your efforts and achieve your goals on a sustainable basis.

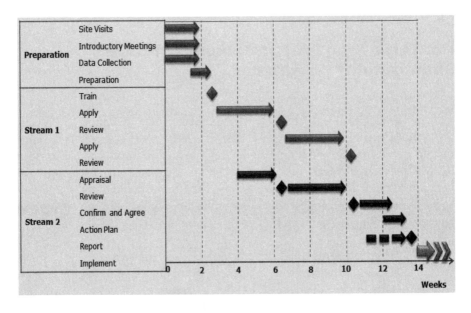

Figure 10-2: The Sequence of Activity for an Inventory Review Program

Preparation Stage

The two keys to a successful program are: having willing participants and collecting good data to show what is possible. The Preparation Stage is where this is established.

Purpose

The purpose of the Preparation Stage is to:

- Ensure that the likely workshop participants and their supervisors/managers understand the goals of the program and the commitment required. The aim here is to generate their 'buy in' so that the program can continue without undue interruption.

- To review the existing inventory and collect data on the inventory holding. The Champion will use this information to help guide the team and show them what is possible in inventory reduction. For example, many people are amazed when shown that the Pareto principle holds true for their inventory. The Champion should be prepared to demonstrate this in the first meeting/workshop.

Who Is Involved

- The Champion
- The likely team members
- Key influencers – the supervisors/managers of the workshop participants

Key Steps and Actions

Site Visits

If possible, visit each location that will be involved in the program. These visits give Champions a feel for the environment, size, standard of housekeeping, quality of processes, level of resources, and busyness for each location. This book is not intended to train Champions in all aspects of inventory management so it is assumed they can recognize the good, the bad, and the ugly when visiting a site.

Introductory Meetings

Introductory meetings with the key influencers give Champions the opportunity to explain the program and its goals and to answer any questions or concerns. Note that the key influencers may not always be the program participants. Often key influencers are the program participant's immediate supervisors or managers. It is these people who influence how individual team members spend their time and how important the program is perceived to be. Champions must work with these people to ensure cooperation and team member focus.

Data Collection

Champions should collect the basic inventory data so that they can conduct their own Pareto review in Stream 1. The success of the program often hinges on the initial data collection by the Champions. Included in this book as Appendix B is a Data Collection Guide that contains 28 questions that will help the Champions to understand the way in which the inventory is managed. This will be most important when considering reduction issues and options plus when determining policy and measurement decisions.

Expected Outcomes

At the conclusion of this stage, champions should be in a position to prepare confidently for the following streams. They will have identified all of the program participants and will have scheduled at least the first team meeting/workshop. They will also have answered questions for the key influencers and have them aligned with the program goals.

Stream 1: The Individual Parts Review

Stream 1 is where the key training is conducted and the use of the material begins. In this stage, the Champions and the team members will also establish their inventory reduction targets.

Purpose

The purpose of the Stream 1 is to:

- Conduct the training so that the team understands the Inventory Cash Release® Process and its application
- Commence the planning and execution of some actions to get early gains. Achieving early gains is important because this will give the program some momentum. Nothing provides encouragement more than success.

Who Is Involved

- The Champion
- The Team Members

Key Steps and Actions

Meeting/Workshop Preparation
This is the opportunity to check that everything is in readiness before the champions get the team together for the first meeting/workshop. Key questions to ask are:

- Is the data organized?
- How is the data to be presented (data projector)?
- Are the materials prepared for the chosen method?
- Is the agenda organized?
- Do the Champions understand the material?
- What will you provide to the attendees by way of handouts?
- Is the venue OK (e.g., is it big enough)?
- Is lunch organized? (Don't laugh; I have seen places where this has been completely forgotten!)

Initial Training
This is where the program participants get exposure to the process and the other content of this book. This training should be divided into two sessions. The first session focuses on reviewing the process and other material. The second session gives everyone a chance to plan how they will actually apply the material. Both sessions may be run in a single day or they may be on separate days — just not too far apart!

At the conclusion of the training you will require two outcomes. First, everyone will have some actions to follow up. Second, the representatives from each location will need to begin applying the step-by-step process to their inventory and identifying opportunities. It is recommended that each location be asked to begin by reviewing the top 250 items on their Pareto list.

Application
Following the initial training, Champions should give all participants a few days to begin their tasks and then follow up with everyone on their progress, hurdles, issues, and so forth. It is the Champions' job to make sure that the Pareto analysis and 7 Actions are being applied correctly and thoroughly. Success relies on objectivity. Watch out for people applying the approach in a way that suits them rather than following the process as laid out herein. Sometimes people will ignore this approach and apply some other process more comfortable to them. This is a trap for those who have not previously experienced this approach. If progress is not being made, your champion may have to follow up more directly.

During this time, champions should also be preparing a master list of opportunities and the likely timing of their impact. Specialist software has been developed to help manage this activity and to track the implementation. Details can be found at the back of this book, at the Web site *www.InitiateAction.com*, or by contacting the author directly.

First Review
This review is essentially an opportunity for everyone to demonstrate their progress. The real value, however, comes from team members seeing the actions of their peers and learning from each other. Conduct this review as a workshop so that any significant hurdles and issues can be identified and corrective actions assigned.

Application 2
Following the first review, the champions should once again give all participants a few days to begin their tasks and then follow up with everyone on their progress, hurdles, and issues. It is at this stage that the champions will need to decide whether to be more involved to help people through the process. It often helps to visit each site again to ensure that any site specific issues and hurdles can be addressed and that the process is fully understood by all participants.

Expected Outcomes

At the conclusion of this stream, each participating location should have a comprehensive list of opportunities and should have made some progress in implementing them. Remember that inventory often takes time to work out of the system, so do not expect overnight results. The master list should also be up to date; you may even be able to forecast the total inventory reduction target that the actions will deliver.

Stream 2: The Management Process review

Stream 2 is where the team undertakes the Management Process Review, creates the implementation plan and establishes the principles, policies, measures, and reporting for ongoing activity and a sustainable result.

Purpose

The purpose of the Stream is to:

- Establish at least a first draft of the principles, policies, measures, and reporting that will enable a sustainable result. Remember that your past inventory outcomes have been a result of the existing principles, policies, measures and reporting. If you are to deliver a different result over the long term, then these will need to change.
- Continue the process of identifying and implementing opportunities.
- Identify the requirements for implementing changes and who needs to be involved in communicating the new approach.

Who Is Involved

- The Champion
- The Team Members

Key Steps and Actions

Appraisal
This step is where you identify any issues and behaviors associated with your current inventory management practices that have a negative impact on inventory levels. To do this you need to undertake both a qualitative review (that is a review of practices, knowledge, and behaviors) and a quantitative review (that is a review of data).

Qualitative Review

- Identify issues such as a misalignment of responsibilities that the categories used for stocking policies are MECE, and whether policies and procedures actually exist.

- Collect and become familiar with all relevant policies and procedures, ensuring that you understand who is responsible for which actions. Do these all make sense? Are there any obvious gaps in policy and procedure? Don't forget accounting policies.

- Develop an interview guide or questionnaire. If you are going to interview a number of people, it is best to develop a set of questions to ensure that you are consistent in your discussions and constantly direct your queries at the key issues. Appendix B contains a 28-question Data Collection Guide that you could use as a starting point. As already mentioned, you should also draw your inspiration from the issues raised previously in this book.

- Visit the Relevant Sites. If possible, visit each location that will be involved in the program. These visits give you a feel for the environment, size, standard of housekeeping, quality of processes, level of resources, and busyness for each location; they enable you to look for anomalies in execution and differences between what people say they do and what they actually do.

Quantitative Review

- Conduct an independent analysis of the relevant inventory data to identify any issues that might not otherwise come to light. This might include:
    - Stock turns
    - Systematic overstocks
    - A large number of items with no stock control limits
    - Stock held above their natural maximum
    - Apparent overuse of just one stock category (i.e., most SKUs are critical)
    - General quality of the data — whether it is up to date
    - Value of recent provisions and write downs

- The patterns and issues that are identified here are those that can only be identified by a global view rather than the specific item view as in the Stream 1 activity. These are generally driven by management and process issues.

- Also try to coordinate what is seen and heard in the Qualitative Review with what is found in the Quantitative Review.

Review of Issues

Before preceding you must work through with the key players on your inventory review team any issues that the independent review identifies. Be sure to engage those who are involved in managing or executing the activity with which there is an issue. Failure to complete this step may result in pursuing issues that result from a misunderstanding and will make the task of building team buy-in of the solution more difficult.

The best way to conduct this review is to view a meeting/workshop with the team members. This meeting/workshop should begin with an update on progress by each of the locations. After this you can move onto discussing issues and then principles, policies, measures, and reporting. It is important to remember that you need to encourage discussion and actually get the participants to participate.

Confirmation

Having reviewed the issues with the personnel involved, you must now confirm and agree on them with the rest of the team, including the supervisors and managers responsible. This may be an iterative process because people often need to see evidence and understand the impact of the issues before they start to take ownership. It is also now time to assign someone the task of creating the first draft of any new principles, policies, measures, and reporting. Relying on a room of people to develop the draft is asking for trouble!

Develop Your Action Plan

Action plans take many and varied forms in different companies and you should use the format that is most familiar to your team.

The champions should by now be able to finalize the master list of opportunities, the likely timing of their impact, and who will be responsible for taking action. It is also a good idea to collate additional actions that may need to be undertaken. For example, additional analysis may be needed or the Pareto analysis extended.

This is the final meeting/workshop in this part of your program so you should now:

- Confirm the list of opportunities
- Review your 'first draft' principles, policies, measures, and reporting
- Review any implementation issues
- Develop your communications plan
- Identify any further actions

Although this is the end of the initial process, it should not be the end of your program. Companies that continue to work on implementation on an ongoing basis achieve greater success than those that don't. Typically this is done by having the location representatives meet on a monthly basis to review their progress and the implementation of their new principles, policies, measures, and reporting.

Report

If this program has been running under the direction of a program sponsor (typically a senior manager), it is always good to provide a written report on the outcomes. Of course, you should keep them up to date during the program, but a final report is always good practice.

Implementation
Many people say that this is where the real work begins. Having reached agreement is one thing, actually getting people to change the way they do things is something else entirely.

Implementation involves two key activities:

1. Problem solving
2. Change management

Problem solving is usually a term associated with an analytical process, but it is also a prime element of implementation. This is because 'the plan' will not be perfect. Often companies try to anticipate every hurdle and every issue. They try to have a plan that will require no correction when commenced—it must work first time. Yet in any journey, there will always be a need to make course corrections.

Problem solving is required as you encounter hurdles and issues that you had not considered. In any implementation, expect problems. Expect issues to arise that you had not thought of. Expect issues to arise that you had thought of, but had not resolved. You can also expect some things that you thought might be issues to not be issues at all!

Change management is the other key activity in implementation and, like problem solving, is a big topic on its own. There have been plenty of books written that deal with this topic extensively; however, in the space available here I am going to present a summary of the issues that you need to understand.

The main point is that change management requires an understanding of the human issues faced by you and your team. That's right; implementation is a human process, not a scheduling process. Planning and scheduling are tools that help manage the human process; they do not replace it.

There are many models of the process for change. They usually include words such as stagnation, grieving, acceptance, relief, reality, resistance, confrontation, critical mass, and point of no return (typically in that order). Because of the human element in change management, these models are based on psychology and the words used often have little meaning in this context to the average manager.

Here is a much simpler approach that works. No matter which change model you subscribe to, the key issues that you need to address are communication, taking action, and reinforcement. Here are my four key actions:

1. Communicate the vision and new process.
2. Take action—actually do the new activities.
3. Reinforce the desired behavior.
4. Manage the tension between the old and the new.

Communicating the vision and the process to achieve that vision requires at least three times more effort than most companies put into this step! Too often companies leap into the 'just do it' approach; the only people who really understand why new things are happening and where this is all going are the ones who wrote the plan. People need to understand where they are going and how they are expected to get there.

Taking action requires you to get others involved and actually start doing something different—preferably executing the actions in your plan.

Once you take action, you must then reinforce the desired behavior. Positive reinforcement is clearly preferred, but behavior correction and follow up also have a place, not in a berating sense, but in a 'that's not the way we do things anymore' sense.

Furthermore, do not underestimate the power of regular review meetings with those assigned responsibility. They provide a forum to air issues that arise. They can be the catalyst for problem solving, an opportunity for positive reinforcement, and a chance for behavior correction. The implementation of the plan must be seen as important and can be achieved by demonstrating the willingness of senior personnel to dedicate time to these meetings.

Measurement systems are a must and keeping score is important. However, you need to track and measure actions, not just outcomes. Let me explain this.

In many situations, outcomes such as reduced inventory can be achieved in a number of ways, including some that are just not sustainable. For example, an apparent reduction in inventory can be achieved by stopping the reordering of fast moving items. This action will produce a short-term result, but not necessarily a sustainable outcome. For a sustainable outcome you should manage implementation by tracking that you have 'done action A' rather than just 'achieved outcome X.'

Change ultimately requires a transition. In turn, this transition creates a tension between the old culture, where people know what to do and are generally comfortable, and the new culture, where there may be uncertainty and people are not yet comfortable. As a rule of thumb, most people prefer what they know. There is less risk and uncertainty.

This tension shows itself as a resistance to change. What you must do in implementing change is manage this tension. That is, understand that it exists, that it is normal, and that the way to relieve the tension is to move people step by step towards the new. The more people understand as they progress, the less fear and uncertainty they will have about the change.

What you **must not do** is allow the old culture to design the new way of doing things. Invariably, this approach manages the tension of change by designing something that is not too different to the current situation or requires everyone else to change (blame shifting).

Expected Outcomes

At the conclusion of this stage you should be executing your actions and predicting your inventory reduction outcome with confidence. It may take a year (or more) to achieve this goal, but you should now be able to project the timing and total quantum of your opportunity. You should also be able to implement the new principles, policies, measures, and reporting that will enable you to sustain the result, and indeed improve upon it, for many years to come.

Measuring Your Progress

One of the key features of successful programs is that you not only measure progress but also report progress in a highly visible manner to the highest level possible. Figure 10-3 shows a sample template that you can use for this purpose.

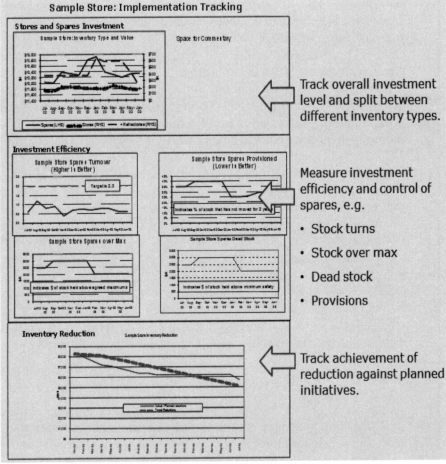

Figure 10-3: Inventory Program Reporting Template

The key elements of this template are:

1. Using charts to track the trends.
   Snapshots of these types of data are rarely of much value as they lack a basis of comparison, so you must use a chart to track the trend. There is a saying in consulting that 'the trend is your friend' and with inventory this is especially so.

2. Measure inventory efficiency and control.
   Measuring just inventory levels is not sufficient as it provides no information about investment efficiency or control. For example, a large reduction in inventory might be impressive, but less so if the stock turn is still very low or comes only from fast movers. (In this case, you would notice that the stock over max or dead stock values won't change.) The measures suggested here are stock turns, stock over max, dead stock, and provisions. You should select measures that you feel are right for your situation.

3. Track achievement against planned initiatives.
   The reductions and trends over time should be compared to the expected impact of the planned initiatives. After all, significant effort has gone into the plan. Variations should be explainable at the initiative level.

Integrating New Inventory Items

For many companies, their inventory really is the forgotten investment. That is, they have forgotten that the inventory is an investment. These companies may expense their inventory or may not have in place any system to manage their inventory. In either case, their issue is not inventory reduction. Instead, these companies are faced with the challenge of setting up an inventory management system. What they need is a way to minimize the likelihood that today's decisions become tomorrow's problems. After all, inventory review is only a way to recognize and correct problems that are now evident from past decisions.

The other challenge, even if you have an inventory management system, is how to decide what to hold when new items are needed. After all, most organizations will add new inventory items every year. This inventory could be new stock lines in sales, new spares for an OEM, new engineering spares for new capital equipment, or just the recognition of a need for something that was not previously included as inventory. The problem is that, when deciding what items to put into inventory, many companies limit their thinking and limited thinking results in being overstocked.

They often ask themselves only one question, 'Does this item need to be available at short notice?' If the answer to this question is 'yes,' they put the item into inventory. The company is then committed not only to the initial expenditure, but also to an ongoing investment. The thinking stops here and administration processes take over.

So how do you use what you have learned in this book to help make better decisions for new inventory?

The starting point is mindset. If you start with a mindset that says availability = reliability or that availability = sales, then you will overinvest in spares because you will justify everything on a false premise.

Your starting point should be to adopt a Zero Inventory Mindset. This does not mean that you hold zero inventory. Instead, it means that you start with a mindset that inventory = cash. You seek to minimize your cash investment without jeopardizing your minimum requirements for availability.

Adopting a zero inventory mindset is especially important because any decision you make on Day 1 is unlikely to be reviewed for some years. Therefore, that decision commits your company to an investment that may not be corrected for years to come. If that investment was for capital equipment rather than working capital, then I am certain that the decision process would be very different. Adopting a sound approach to these early inventory decisions may prevent the need for future rationalization!

After asking whether an item needs to be available at short notice, two further questions should be asked before any item is taken up in inventory. First, do *we* have to make the investment? Second, if so, how can we *minimize* the investment?

In asking both of these questions, you are seeking to be smart about how you spend and invest your cash.

Asking 'Do *we* have to make the investment?' seeks to avoid making any investment by either finding a substitute or getting someone else to make the investment on a consignment basis. A consignment arrangement is always the preferred option if demand can be satisfied at an acceptable service level.

If no one else is prepared to invest or there are no substitutes, then you must ask 'How can we *minimize* the investment?' In other words, how can you spend as little of your cash as possible and still have the items available? The focus is now on setting up the dynamics of supply to minimize the cash you need to invest. Options include matching delivery with usage, minimizing the reorder quantity, reducing the replenishment time, or simply buying the lowest total cost item. These are the inventory reduction actions associated with 'taking more items out.'

Two checklists are included at the end of the chapter to help you make better Day 1 decisions. The first, Figure 10-4, is for engineering inventory and the second, Figure 10-5, is for general inventory items. The key difference is in the sub-questions that you ask rather than the three key questions discussed above.

These checklists are not meant to be perfect for all situations; they are meant to guide you in the development of your own checklist. What you can then do is make that checklist a compulsory part of your process for adding new inventory items. Using a checklist doesn't ensure that mistakes won't be made. However, it does provoke the thinking of your team, helps establish that Zero Inventory Mindset, and establishes achieving true inventory optimization as a part of your culture.

Integrating Inventory Process Optimization™ with Inventory Management

Successful and sustainable inventory reduction is as much a result of the culture relating to inventory in your organization as it is a result of applying the right process. Culture is the way that your company does business. It is the attitude towards inventory that the company as a whole adopts. In a sales organization, is inventory measured in terms of the margin that it can produce? In maintenance, is inventory measured by availability? In both of these examples, the culture leads to an over-investment in inventory.

The purpose of reviewing policies, processes, measures, and reporting is to be able to influence the culture over an extended period of time (it won't change overnight). If you limit your inventory review program to a project that lasts just the 14 weeks of training and initial application you will send a signal that inventory review is something that happens every now and then, but not as a part of 'how we do business'. The challenge is to be able to integrate your inventory review actions with your ongoing inventory management. To do this, you need to be sure that some actions become part of your regular ongoing activity and it is suggested that:

- Your team complete a Pareto review on an annual basis.
- All items of inventory be reviewed within the time frame before your accounting system says that they should begin being written down. The timeframe for reviewing all items will therefore depend on your local tax rules and company policy.
- New items being added to inventory need to be reviewed, as per the process discussed in the previous section.
- Your reporting on inventory needs to continue to reinforce the processes that you are now putting in place.

In short, inventory review should not be thought of as a project that is completed after 14 weeks. Inventory review is a culture—an attitude—and it needs to integrate with your ongoing inventory management and become a part of the way you do business.

Questions to Ask	Yes	No
1. Must the item actually be stocked?		
If the failure occurs is there an alternative which does not jeopardize: • Safety • The environment • The supply promise • Quality		
Could the potential failure be detected and managed in a cost efficient manner before actual failure occurs?		
Could the item be repaired in a suitable timeframe?		
Could the item already be supplied rapidly by a local vendor?		
Can you use something that you already have? • Substitution • Duplication • Rationalization		
If you answer yes to any of the above, then follow that option before progressing.		
2. Will someone else make the investment?		
Can you get someone else to pay for the item? • Consignment • VMI • Shared ownership/pooling		
If you answer yes to this question, then follow that option before progressing.		
3. How can you minimize the investment?		
Review the factors that drive safety stock.		
Reduce the reorder quantity.		
Match delivery with usage.		
Reduce the value of the item.		
If you answer yes to any of the above, then follow that option when creating a new stock item.		

Figure 10-4: The Stock Decision Checklist For Engineering Inventory

Questions to Ask	Yes	No
1. Must the item actually be stocked?		
If there is demand but no stock, will the customer wait for the item?		
Could the item already be supplied rapidly by a local vendor?		
Can you use something that you already have? • Substitution • Duplication • Rationalization		
If you answer yes to any of the above, then follow that option before progressing.		
2. Will someone else make the investment?		
Can you get someone else to pay for the item? • Consignment • VMI • Shared ownership/pooling		
If you answer yes to this question, then follow that option before progressing.		
3. How can you minimize the investment?		
Review the factors that drive safety stock.		
Reduce the reorder quantity.		
Match delivery with usage.		
Reduce the value of the item.		
If you answer yes to any of the above, then follow that option when creating a new stock item.		

Figure 10-5: The Stock Decision Checklist For General Inventory

Chapter 11

Case Studies

This chapter contains two case studies that demonstrate the application of several different processes and concepts discussed within this book. The case studies are presented here in simple formats that progress from background information, to the process undertaken, to findings and issues, and then results. The details are shown with a combination of commentary, charts, and figures relating to key aspects of each case. The charts and figures included are, for the most part, those that were used by the companies themselves while undertaking and reporting their programs. However, information that might specifically identify the companies has been deleted.

The first case study, *Establishing a Materials and Inventory Management System*, addresses the important issue of people and processes in materials and inventory management. This covers the materials discussed in Chapters 4, 5, 6, and 7. In this case study, the company had recently taken ownership of another company and that company had little in the way of established policies and procedures. In order to move them along the path towards achieving smart inventory solutions, they needed to develop and implement a full range of policies and procedures, including:

- Corporate level policy requiring board-level approval
- Operational procedures for use on a daily basis
- Stocking policies
- Storeroom organization guidelines
- Storage guidelines

The second case study, *Applying the Inventory Process Optimization™ Method*, shows the application of the Inventory Process Optimization™ Method in the steel industry. Although the method and related concepts have been applied in almost all parts of the steel industry manufacturing and delivery supply chain two example locations are chosen for this demonstration. Similarly, the method has been successfully applied across many different industries but, due to space limitations, just the steel industry was chosen for inclusion here. This case study uses the method described in Chapter 9 as well as the material discussed in Chapters 6, 7, and 10.

There are three important issues raised by this case study:

1. The Pareto Principle holds firm for all inventories — at least the ones that I have seen!
2. The management issues are similar across most businesses and you will most probably see your own issues in the lists shown.
3. The full results are achieved over time. The case study shows results at both the conclusion of the initial consulting program and the results that were achieved through continued application of the process over several years.

Case Study 1: Establishing a Materials and Inventory Management System

Background

This case study shows the process applied to establish policies and procedures for materials and inventory management and the issues faced during program implementation. The company concerned, which had taken over another organization, needed to ensure appropriate management and governance of their inventory investment across 25 individual locations.

The key issues they faced included:

- No reliable records of the quantity, quality, or location of spares held
- No tracking of usage
- No standards for storage
- No identification of critical items
- Poor control of repairable spares
- Imminent implementation of a new Computerized Maintenance Management System (CMMS)

They needed a materials inventory management system that provided appropriate control and integrated with the new CMMS. Key questions for this company were:

1. Are current inventory arrangements appropriate to the types and value of the inventory?
2. What actions, if any, could be taken to improve the management and governance of the inventory investment?

The Process

Following an initial Snapshot Review, a four-stage process was designed and implemented. A Snapshot Review evaluates capability, accountability, and systems, showing how and where a company needs to improve in order to achieve its goals. The process involves both qualitative and quantitative analyses to ensure that all aspects of inventory management are covered. The four-stage process was designed to provide every opportunity for 'buy in' by all staff during the implementation (Figure 11-1).

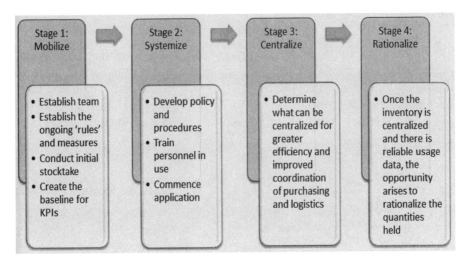

Figure 11-1: Four-Stage Implementation Process

The key activities included the following steps:
- Establish a steering committee.
- Review governing accounting standards and their implications.
- Manage internal communications on the project progress and process.
- Develop stocktake processes and documentation.
- Conduct a pilot stocktake, followed by a company-wide stocktake.
- Visit a 'best practice' site to review its infrastructure and practices.
- Develop standards for materials storage and storeroom/yard layout.
- Coordinate clean-up of locations
- Develop and document new policies and procedures.
- Approve policies and procedures at board and executive level.
- Review infrastructure needs and develop an initial budget for infrastructure improvements.
- Review CMMS capability and implementation issues.
- Conduct awareness training for key personnel.

Issues Faced During Implementation

During the development and implementation of the policies and procedures, a number of relatively common issues were faced. Table 11-1 summarizes these and the mitigation actions.

Table 11-1: Issues and Mitigation Actions

Issue	Mitigation
Lack of 'buy in' by middle management.	• Maintain Senior Management support and develop support at the 'coal face' (working around the mid level). • Continue communication with mid-level management, including demonstrating benefits. • Establish auditable procedures and standards. • Establish personal implementation-related KPIs for key managers.
During the program, the initial champion for materials management moved on in the organization.	• Executive level 'buy in' used to continue the push of the program. • Local champions established at each location as people showed interest.
Little organizational understanding of 'good practice'.	• Visited an external 'good practice' location to demonstrate what 'good practice' looked like and that it was possible to achieve in their industry. • Development of written standards which included (where appropriate) photos of good practice. • Establishment of an internal 'good practice' location that was used as a role model for other locations.

Insufficient data to establish reordering parameters.	• Development of a decision making model which used known information to establish the most appropriate calculation approach and then provided a sensitivity analysis to model the effect of different estimates of requirements.
Need to establish ordering parameters for a large number of items.	• Establish a Materials Usage Review Group to identify critical items and ensure sufficient procurement while data for automated reordering was being established.
Lack of trust that the new system will ensure availability of the required materials.	• Materials Usage Review Group monitored levels while trust in the system was established. • Issues and concerns relating to trust were proactively identified and followed up.
Inadequate storage infrastructure	• Established an infrastructure plan with the aim to gain funding approval and establish the required infrastructure over several years.

Table 11-1: Issues and Mitigation Actions (continued)

The development of policies and procedures was critical to the program's success. Figure 11-2 shows how 'swimming pool' charts help simplify the presentation of procedures and clarify the responsibility and sequence for each action.

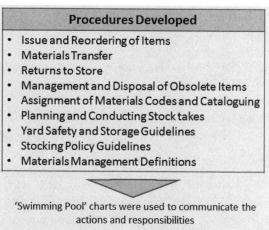

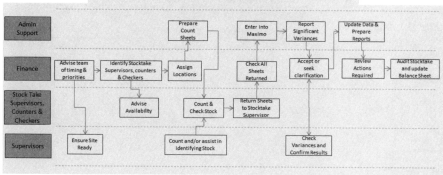

Figure 11-2: Communicating Procedures

Results

At the time of writing, the program was still underway. However, the appropriate systems for management and governance of inventory investment had been established, including:

- Materials Management Policy and Procedures developed and implemented.
- Systems established for recording and tracking materials usage.
- Procedures established for conducting stock takes.
- Standards established for labeling and storage.
- Guidelines established to identify critical and other spares types.
- Guidelines established for establishing stock levels and reordering of materials.
- Guidelines established for improved control of repairable spares.
- CMMS implementation coordinated with materials management improvements.
- Full transparency of stock holdings at all locations established.
- More than 100 staff trained in application of new policies and procedures.

By a) tackling their issues relating to people, processes, policies, and procedures, b) identifying and addressing internal myths, and c) formalizing their approach to managing the spare parts store room, this organization has set itself up to achieve long-term success with its materials and engineering spares management.

Case Study 2: Applying the Inventory Process Optimization™ Method

Background

The Inventory Process Optimization™ method has been applied in many segments of steel production and processing, including:

- Blast furnace
- BOS steelmaking
- Coke ovens
- EAF 'Mini Mill'
- Billet casters
- Slab casters
- Structural steel mills
- Rod and bar mills
- Wire mills
- Pipe and tube mills

This case study includes results from two sites.

The Process

In each case the process applied was as described in this book, including the 14-week program of training and initial implementation discussed in Chapter 10, *Execution: Taking Action to Achieve Results*. A summary of the method is shown in Figure 11-3.

Review teams included people from the different disciplines that have an impact on inventory holdings:

- Maintenance
- Operations
- Procurement/Supply
- Storeroom/Spares Officers
- Engineering
- Finance

The mix of personnel included people from both the shop floor and management. Local champions were appointed to ensure ownership of the process and actions.

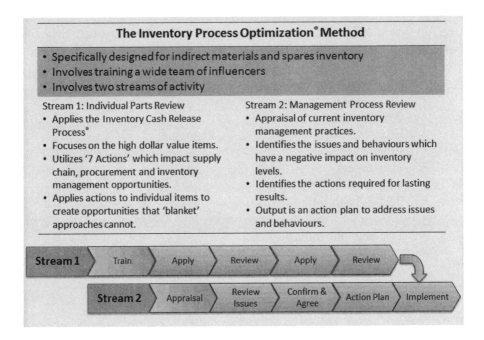

Figure 11-3: The Process Implemented

Findings

In all cases, the Pareto Principle held true as a majority of inventory value was tied up in a small number of items. For the two sites in this case study, the effect was even more pronounced (Figure 11-4):

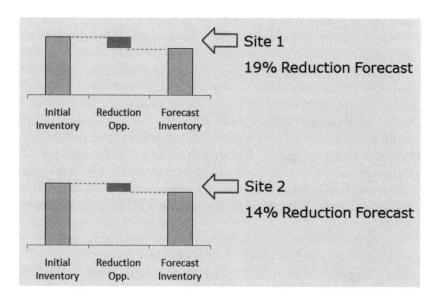

Figure 11-4: Pareto Results for Two Locations

The management reviews identified a range of inventory management issues (Table 11-2):

Site 1	Site 2
• Conflicting priorities with plant personnel • Lack of ownership/accountability • No consequences for those making stocking decisions • No financial measure of the benefits • No clear reporting structure around spares • Production/availability focus • Too much access to charge out stock • Poor risk assessment • Setting of max-mins — wider input required • Reorder quantities too high • System forces order of parts not yet required • Perception that supplier lead time is longer than it actually is • Lead times not revisited	• No history on usage • Initial stock decisions not revisited • Suspect master data – especially lead times and splitting repair items from new • System drives long lead times — all suppliers • Process drives the non identification of obsolete items — if item made obsolete then it hits the R&M budget • Alignment of responsibilities — financial responsibility vs. setting stock level • Plant reliability drives breakdowns and the need for emergency spares • Risk assessment is non-existent

Table 11-2: Management Issues Identified

Results

Selected KPIs were measured using an inventory tracking 'dashboard' that was reported to an executive level (Figure 11-5).

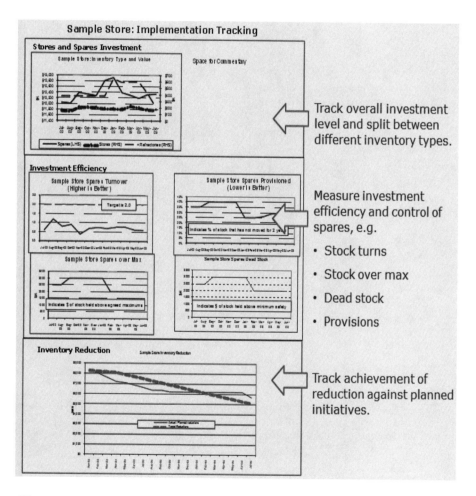

Figure 11-5: Example of Inventory Tracking 'Dashboard'

Initial Results

During the programs, the teams identified acceptable but not aggressive levels of inventory reduction (Figure 11-6).

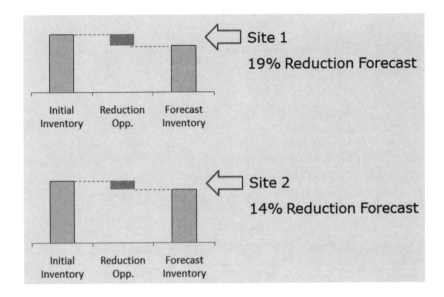

Figure 11-6: Initial Results

Final Results

However, by continuing to apply the method over time, the teams delivered significantly greater (and sustainable) reductions (Figure 11-7).

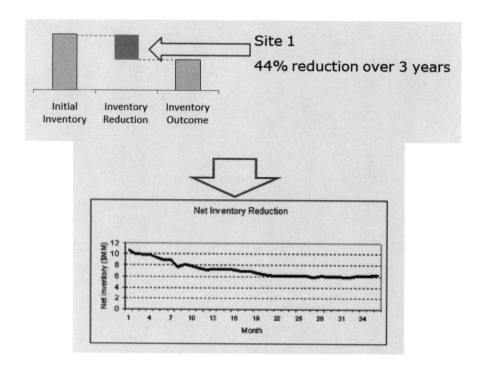

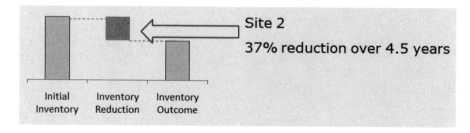

Figure 11-7: Results Achieved Over Time

Chapter 12

Where To From Here?

Sydney Australia is perfect for a picture postcard. The natural beauty of the harbor is complemented by the design brilliance of the opera house and this is all framed by the majestic Sydney Harbor Bridge (which is affectionately referred to locally as the 'coat hanger' because of its steel arch construction).

As you drive west from this famous scene, the landscape changes into a fairly generic hinterland reminiscent of the industry that surrounds almost every major city. Sydney's industrial hinterland includes oil and gas refineries, steel mills, paper plants, and major manufacturing locations for global brand names producing everything from biscuits to chemicals, glass, pharmaceuticals, valves, and everything in between. As you can imagine, I have visited this region often for my consulting work.

One clear and cold spring morning, I parked my rental car in the car park of a typical western Sydney facility. This facility was a small part of a very large company with global operations. The CEO of that company had engaged me to help the organization achieve true inventory optimization; this stop was one of two dozen similar facilities that I was in the process of visiting. My appointment was with the site's Manager of Operations and Maintenance — the most senior onsite engineer. On walking into reception, I signed in as a visitor and the receptionist told me that 'they' were waiting for me in the training room in the next building. She asked if I would mind finding my own way there. At this point I sensed that this was not going to be a typical first meeting.

Finding the training room, I knocked on the door and entered. The room was a standard corporate training room with a projector screen and white board at one end, a projector suspended from the ceiling in the middle, and under that a series of tables all joined together to form one long table. What was less typical was that my contact and all of his direct reports, all ten of them, were seated on one side of the table, facing the door, with their arms crossed! It looked and felt like I had walked into a panel style interrogation.

Of course everyone was quite civil; Australians are friendly in that way. However, the attitude and position were clear — we don't want you here, we don't need you here, we don't have any problems with our engineering spare parts. I asked why they were so certain of this and the answer was universal: They almost never had a stockout and, therefore, in their view, they didn't have a problem. As you will see, they had fallen into the Service Factor Trap.

Theirs was a textbook example of the biases that are described in this book (page 52) and which represent one of the greatest hurdles in implementing a successful inventory review program. This group, which meant well, was demonstrating almost every bias discussed. The members:

- Were overconfident in the ability of their systems to provide true optimization.
- Believed that bringing in outside assistance was a loss of face on their part and showed a lack of confidence in them by the head office.
- Framed their options by only looking at solutions relating to parts supply, and not inventory investment.
- Anchored their beliefs in the idea that a stockout was the only measure that counted.
- Were concerned about any changes that would alter the status quo.

The only way to overcome these biases was to forge ahead with the agreed program so that they could discover for themselves what they needed to do. Here's some of what we found using the data collection guide found in Appendix C:

- Stockouts were not actually measured!
- Only one person had any type of stores or inventory performance measure (which was the stockout measure) but, as mentioned, this was not actually measured.
- No one was responsible for the level of financial investment in inventory.
- Their stock turn ratio was 1.1; this level was below the stock turn average of 1.25 across their sister facilities.

- Although there was a main central store, there were seven official but uncontrolled other stores across the site.
- There were no standard definitions to classify items or from which to set stocking policies.
- In their own words, 'not a lot of science' went into setting the ROP and ROQ for new items. Even then, it was 'set and forget, unless we run out'.
- The stock control function of their ERP was not turned on.
- Obsolescence was managed 'if someone happens to notice'. Notifying the Storeroom of newly obsolete items was officially the responsibility of any engineer uninstalling equipment but it was not normally done.
- Their storeroom personnel failed to receipt items consistently, if at all.
- There were no storeroom KPIs.
- The central storeroom had poor housekeeping with items routinely stored in the aisle, making it difficult to move around.

Once Stream 1 of the program started and the training commenced, the manager of Operations and Maintenance — to his credit — immediately asked me to present another workshop involving a wider group of personnel as well as their onsite executive committee. He had realized that there were actually many issues and that the task was a bigger one, involving more than just his team. Subsequently:

- Write downs for the preceding 12 months were identified to be approximately 9% of the inventory value (based on an existing accounting policy).
- Their Pareto review showed that 70% of their dollar value was tied up in less than 5% of their inventory, meaning that they actually needed to focus on just 500 items to have a major impact.
- Reviewing those 500 items identified:
  - A single opportunity valued at $425,000 from just one SKU, where the stock level had not been adjusted following reduced usage.
  - Another single SKU with $200,000 in stock was identified as obsolete and a further $200,000 in outstanding orders for that item were cancelled. (Yes! An obsolete item had been reordered!)
  - A single SKU where $50,000 had been ordered 'by accident'.

- A further 45 individual opportunities to reduce inventory by more than $10,000 each (some worth as much as $70,000).
- It was estimated that 30% of the items removed from their central storeroom were not officially signed out. Therefore, they had no official record of their usage and their existing records were unreliable.

After working through the program described herein, their spares and stores holding fell by 45%, over a period of 16 months. They then remained at that level for the next two years. Accordingly, their stock turn went from 1.1 to 1.9 over the same period.

Although the inventory reduction is impressive enough, the fact that it remained low over the subsequent years demonstrates that they were not adversely affected by this reduction. In fact, it is safe to say that prior to this program they had in effect spent millions of dollars on items that they really didn't need to support their operational outcomes.

So what does this mean to you?

First, the ideas, tools, and processes discussed in this book work. They just do. If you are serious about achieving true inventory optimization, then apply these ideas, tools, and processes to your organization. Don't cherry pick those that you like or agree with — stick with the program and results will follow.

Second, take action. Don't be delayed or deflected by discussions and arguments that 'everything is OK'. It isn't. The ideas, tools, and processes in this book have been applied in plants and facilities of all different types and sizes all over the world. From oil and gas processing to steel, aluminum, and metals processes; to power generation and other utilities; to railways and transport. From industrial spare parts to titanium hip joints. From the Middle East to Asia and the Pacific to the Americas. You are not so unique that this will not work for you. Every single place has at least one person who says 'nothing needs to change,' yet I have yet to see one single application of the ideas, tools, and processes in this book that has not delivered significant benefits for the organization applying them.

Third, begin today. You have invested time and energy to read this book and learn about achieving true inventory optimization; why not get on with it? Even if you cannot get a full program going, start influencing the people who can collect information, produce your Pareto curve, review storeroom management, use the data collection guide, build a picture of the issues and hurdles you face, identify and quantify the potential benefits, ask questions, dispel myths, and challenge assumptions. Just start.

Implementing smart inventory solutions to achieve true inventory optimization is a journey that is personally satisfying. But more than that, it is a journey that can be truly valuable to your organization. Get on with it.

Bibliography

The following are books and articles that I have found to be helpful in developing the material in this book. In some cases the book or article is directly referenced.

A Decision Making Framework for Managing Maintenance Spare Parts; S. Cavalieri, M. Garetti, M. Macchi, and R Pinto; Production Planning and Control; 19:4 379-396, 2008.

The Concise Oxford Dictionary, Edited by J.B Sykes, Oxford University Press, 1982.

The Fifth Discipline, Peter Senge, Random House, 1992.

Issues in Financial Accounting, Fifth Edition, Scott Henderson and Graham Peirson, Longman Cheshire, 1992.

Lean Thinking, James Womack and Daniel Jones, Simon and Shuster, 1996.

The Machine That Changed the World; James Womack, Daniel Jones, and Daniel Roos, HarperCollins, 1991.

Maintenance Management, Revised Edition, Lawrence Mann, Jr., Lexington Books, 1982.

MRO Inventory Reduction — Challenges and Management: A Case Study of the Tennessee Valley Authority, G.J. Bailey and M.M. Helms, Production Planning and Control, 18:3 261-270, 2007.

Probability and Statistics for Engineers and Scientists, Second Edition, Walpole and Myers, 1978, Macmillan Publishing.

Toyota Production System, Taiichi Ohno, Productivity Press, 1988.

Appendix A
Materials and Inventory
Management Glossary

A

ABC Analysis

An ABC analysis divides the inventory items into categories and helps identify which items are more important. Typically A is more important than B and B is more important than C. For most inventories, the criterion for the division is stock turn, but this is of little value for ABC analysis of MRO and engineering inventory.

Accounting Standards

A regulation that governs the way in which accounting and financial data are to be interpreted with respect to a company's financial position.

Active Stock

Inventory items with regular demand/usage.

Aged Item

An item that has had no demand/usage for a predefined period. In engineering spares, this could be as long as four years.

Automated Ordering

Computer-generated purchasing that is triggered by stock levels and requires no human intervention.

B	

Balance Sheet   A financial report that shows the relative status of a company's assets, liabilities, and shareholder equity. Sometimes now referred to as the Statement of Financial Position.

Bandwidth   The capacity of a computing system to transfer data.

Batch size   A minimum quantity in which an item is made and/or delivered.

Bottleneck   A constraint on process flow. Typically the point at which things happen most slowly.

Brainstorm   A process of capturing ideas without questioning their value in advance.

Breakthrough   The recognition of a change that needs to occur in order to effect an improvement.

Buffer   A reserve of stock that enables continued supply when the timing of supply and demand are mismatched.

Business Environment   The economic and social conditions within which a business operates.

C	

Capital Investment   An outlay of money to acquire or improve a capital asset such as equipment or buildings.

Capital Spares   Spares purchased with a new capital investment such as plant and equipment. Typically carried at nil value as expense is capitalized with the investment.

Cash Flow   A measure of the accumulation or expenditure

of cash as a result of a company doing business. A positive cash flow means that more cash came in than went out; a negative cash flow means that more cash went out than came in.

Circumstantially Overstocked	Overstock that occurs due to a particular set of events or circumstances rather than the specific systems that may be in place.
Consignment Inventory	System where the vendor owns the inventory, but we hold it for our use. The vendor charges for the inventory only when the item is used or consumed.
Consumable Spare	A spare part that is worn out or otherwise used up in the environment in which it is used.
Continuous Improvement	Creating incremental changes that improve the operation or business outcomes.
Critical Inventory	Spares expected to be used in the normal operation of plant and equipment which, if unavailable, would prevent the plant from operating.
Culture	The prevailing behaviors that determine how a company's people respond in particular circumstances, e.g., a proactive culture, a reactive culture.
Cumulative Value	The value of all items on a list taken from the beginning to the point at which the cumulative value is determined.
Current Assets	Assets such as cash, accounts receivable, and inventory that are likely to be able to be converted to cash within a year.
Customer Service	A measure of success in meeting a customer

Level	supply promise. Usually measured in % as in 99% of the time we meet the customer supply promise.
Customer Supply Promise	The implicit or explicit promise made that a supplier will provide goods or services within a certain time frame.

D

Dead Stock	The component of an active inventory item that has no movement over a predefined period, typically two years. This period must be long enough to account for the full range of volatility in demand and supply.

For example, an item may have regular usage and replenishment, but has never had fewer than 3 items in stock over two years. These 3 items are effectively dead stock as they are unlikely to be used at any time. |
Demand	A customer order that requires an item to be supplied from inventory.
Depreciable Spares	A spare that is unique to a specific asset and is depreciated over the life of the asset. Sometimes called a Unique Spare.
Depreciation	A non-cash accounting entry that allows a company to spread the cost of an item across its useful life.
Distribution Requirements Planning	An approach to managing distribution that is similar to Materials Requirements Planning. Essentially the same logic is used, but it is applied to a series of store or warehouse locations.
Dividends	Payment of company earnings to stockholders.

Dollar Value Held	The total investment made in a single line of inventory. Usually the number of items multiplied by the value per item.
Dollar Value Invested	See *Dollar Value Held.*
DRP	See *Distribution Requirements Planning.*

E

Economic Order Quantity (EOQ)	The purchase quantity that represents the minimum cost point between spare parts holding and purchasing costs. EOQ is often misunderstood to be the quantity providing the lowest purchase cost per unit; as a result, many organizations actually overstock their inventory. However, when costs such as obsolescence are taken into account the best purchase quantity may not be that with the lowest purchase cost per unit.
Electronic Delivery Acceptance	The use of computerized systems to process and acknowledge a delivery of goods.
Electronic Funds Transfer	The use of computerized systems to transfer money from one bank account to another.
Engineering Spares	Items held to assist in repairing equipment if it breaks down.
EOQ	See *Economic Order Quantity.*
Excess Stock	There are typically two uses for this term: 1) Inventory that is in excess of the amount needed to cover a predefined period, typically two years, or 2) a level of inventory that is greater than the predetermined maximum.
Expediting	Taking action outside of the normal routine and priority to make sure that goods are delivered

F	
False Economy	An action that appears to provide an economic benefit, but which when looked at broadly does not provide a benefit.
Finished Goods	Items that require no further processing before being dispatched to a customer.
Forecast Accuracy	The degree to which an actual outcome matches a prediction of that outcome. Typically used when predicting sales or demand.
Forecasting	Predicting an outcome. Typically used for sales or demand.

G	
Gross Margin	The net revenue from sales minus the cost of sales. This excludes overhead costs.

H	
High Turn Over	Items that have a stock turn, that is, at least twice the average for the inventory.
Holding Cost	A term used to identify the annual percentage cost of owning inventory. See also *Total Cost Ratio*.
Human Dynamics	The normal behavioral characteristics of people.

I	
ICR®	See *Inventory Cash Release*®
Implementation	Changing management activity and processes so that a different outcome results.

Indirect Inventory	This term covers all inventories that are not used directly in production. This includes all 'bought in' inventories such as parts and components, finished goods, service parts, OEM spares, MRO inventory, engineering spares and industrial supplies. Sometimes referred to as Indirect Stock.
Individual Parts Review	Stream 1 of the Inventory Process Optimization™ Method. Uses the Pareto Principle to focus on the items with the greatest value tied up in inventory and then applies supply chain and process review principles to what is commonly thought of in either mathematical or technical terms.
Insurance Spares	An item that would not be expected to be used in the normal life of plant and equipment but, if not available when needed, would result in significant losses.
Interest	Payment to a lender to compensate for the funds borrowed. Usually measured as a percentage of the funds borrowed.
Inventory	Items that are held in the expectation of future sale or use. Also referred to as stores and stock.
Inventory Cash Release®	A structured step-by-step inventory review technique that guides users through the application of different management methods to match the method with the inventory and so produce the most optimal inventory result. The ICR® technique is supported by a purpose-built software program provided by OMCS International.
Inventory Management	The processes used to control, issue, and order inventory.

Inventory Management X-Ray	The Inventory Management X-Ray is a diagnostic tool that enables an objective review and evaluation of inventory management approaches and outcomes. The tool examines 52 attributes of inventory performance in 8 different areas of inventory management, then builds an objective profile of the inventory management. By comparing the current approach to four predetermined levels of performance, the tool can pin point both strengths and weaknesses.
Inventory Process Optimization™	Inventory Process Optimization™ is a two-stream process that has been specifically designed for application with MRO and engineering spares. The technique involves an individual parts review using the Inventory Cash Release® (ICR®) technique and a management process review that identifies the factors and behaviors that have a negative impact on inventory levels.
Inventory Risk	The risk that an item in inventory may spoil or become obsolete.
Investment	An outlay of money to acquire an asset. See also *Capital Investment*.
Issue Cost	The price that is charged to a cost center when an item is released from inventory

J

JIT	See *Just In Time*.
Just in Time (JIT)	Just In Time is perhaps the term most often quoted in inventory management. In short, JIT seeks to ensure that items are made available only at the time when they are needed. The approach minimizes or eliminates waiting time

and can significantly reduce or even eliminate inventory holdings. Unfortunately it is misunderstood to be a technique when it is really a philosophy and an outcome that results from the application of other techniques (such as the use of Kan Bans) and the development of highly responsive supply lines.

K

Kan Ban

Kan Ban is perhaps the term second most commonly used in inventory management (after Just in Time). A Kan Ban is simply a signal that more of an item is required. It could be in the form of an order ticket or a bin to be filled. A Kan Ban ensures that items are supplied only when they are needed and works best in production environments where the stock is used quickly. In spares management, a two-bin system may be considered to be a Kan Ban, but it would not guarantee the lowest inventory level.

L

Lead Time

The time it takes to replenish an item. Typically measured from the time the trigger point is reached to the time the item is physically restocked.

Lean Thinking

Originally developed as a way of embedding higher quality and lower costs in manufacturing organizations, Lean principles have since been applied in many different organizations and management situations. One of these is the Engineering Materials and Spares Storeroom.

Logistics

The organization and management of the movement of goods.

Low Turnover	Inventory items that have a stock turn that is less than one quarter of the average.

M

Maintenance System	A management approach that aims to ensure that production equipment achieves a predetermined level of availability.
Management Process Review	This review involves understanding your current management practices and behaviors and identifying those that have a negative impact on your inventory holdings.
Materials Resource Planning	MRP — a technique that is applicable to items that have dependant demand, that is, when the quantity required for an item is dependent upon the quantity of another item. For example, if we were making cars we would need to ensure that we had 4 wheels for every car and this could be set up in a plan. MRPs have little use in MRO and engineering inventory except perhaps in kitting situations.
Max–Min	An approach to inventory management that aims to maintain the quantity in inventory between a minimum and maximum level.
Maximum	The maximum quantity of an item that is expected to be held in inventory. The maximum is also used to determine the reorder quantity when the minimum is reached.
Minimum	In a max-min system, the quantity in stock at which replenishment is ordered. In some cases, this is equal to the safety stock quantity. See also *Reorder Point* and *Trigger Point*.
Minimum Order Quantity	The lowest quantity of an item that a supplier will supply.

Momentum	In this instance, refers to an organization having the energy, focus, and direction to ensure that a series of actions is implemented.
Monte Carlo Simulation	An analytical technique that uses random numbers as input variables and applies them to a known function (or formula) to identify possible outcomes. It is named after the random inputs that occur in table games, such as roulette, at the casinos in Monte Carlo.
MRO	Maintenance Repairs and Operations. This term refers to the inventory that is not converted to finished goods in a manufacturing process. It includes spare parts and consumables.
MRP	See *Materials Resource Planning*.

N

Non-Cash Cost	A charge against the Profit & Loss (P&L) that is an accounting entry rather than an expenditure of cash.
Non-Moving Stock	Inventory that has no demand or issue over a predefined period of time.
Non-Trading Stock	All inventory that is used to support production but is not ultimately for sale, e.g., spares and stores. Sometimes referred to as *Indirect Stock*.
Non Value Adding	An action or investment that does not result in an improved outcome as a result of the action or investment.

O

Obsolete	An item that is no longer required for its original purpose.

Obsolete Stock	Any inventory that will no longer be used as a result of plant closure, product replacement, or end of life.
Operating Stock	All raw materials, WIP, and finished goods that are intended for production and sale. Sometimes referred to as *Trading Stock*.
Operational Budget	The financial plan for a company's production or service delivery function.
Optimization	Optimization is perhaps the most misused term in inventory management. Optimization is the tradeoff of one condition against another to find the best position. It is usually incorrectly assumed that optimization is absolute and cannot be improved. This is not the case as improvements can be achieved if the elements being optimized are changed. This is why process optimization is the key to improved inventory performance.
Overheads	The costs of a business that are not directly associated with the manufacture and sale of goods or services.

P

Pareto Principle	Literally, a minority of inputs produces a majority of results. Sometimes referred to as the 80:20 rule.
Per Annum	Per year.
Phantom Stock	Inventory that appears on the official record, but which does not actually exist. Usually results from poor record keeping.
Planned Inventory	Inventory that is not usually held, but which is accumulated for a special project or event.

Policies	Course of action adopted to address an issue based on the rules of conduct or principle.
Principles	Rules of conduct that define how we will behave or respond.
Probsolete Stock	Any stock that is suspected of being obsolete, but which requires further investigation before obsolescence is confirmed.
Procedures	A way of performing a task; a series of actions conducted in a certain order or manner.
Product Complexity	Applies where there may be a large number of variations in style or type for a single basic product.
Profit and Loss (P&L)	A financial report that summarizes a company's revenues and costs. This report can include non-cash costs and is not a report of cash flow. This sometimes now referred to as the Statement of Financial Performance.
Project Budget	The financial plan for a special activity that is not usually part of the normal operating expenditure.
Project stock	Inventory that is accumulated for a particular project or event.
Purchase Order	A document gives authority to a supplier to supply goods and invoice for them.
Purchasing Efficiency	An approach aimed at minimizing the paperwork or administrative activity associated with purchasing or procurement.

R

Range Complexity	Occurs when there are a large number of different items in inventory.

Raw Materials	Materials from which products are made.
RCS	See *Reliability Centered Spares*.
Reliability Centered Spares	Reliability Centered Spares is a technique for determining the appropriate spares level based on expected performance of a reliability program. The approach asks several questions to guide the thinking. The major weaknesses are that it requires significant data and does not seek to address the many factors that influence inventory outcomes. However, if the data is available, it can be useful for setting initial targets and these can then be improved through the use of process optimization.
Reorder Point	The level of inventory at which replenishment is triggered. Also referred to as the *trigger point*. See also *Minimum*.
Reorder Quantity	The predetermined quantity to be reordered when the Reorder Point is reached.
Replenishment	The act of ordering and physically receiving inventory.
Replenishment Cycle	The activity involved in identifying that stock is low, ordering the stock, and making the new stock available for use.
Reservation Systems	A process that enables an item to be held for use only by the person who makes the reservation.
Reward System	The process by which people are rewarded for achievement or positive outcomes.
Risk	Exposure to the chance that something will be lost as a result of some action or inaction.
ROP	See *Reorder Point*.

ROQ	See *Reorder Quantity*.
Rule of Thumb	A rule that approximates an outcome, but generally cannot be proven to be accurate.

S

Safety Stock	Items held as an allowance for mismatches between demand and supply during the lead time to restock.
Salvaged Stock	Inventory that is retained for potential cannibalization, but where any other potential use/sale is unlikely. Typically held at nil value.
Service Level	The service level is the percentage of times that a request for an item is filled in an acceptable time frame; e.g., a 95% service level means that 5% of the time the request is not fulfilled.
SKU	Stock Keeping Unit - Refers to any individual inventory item
Slow Moving Inventory	Inventory that has had no use/consumption for a predefined time frame. In many cases, this time frame is 4 years; however, this may vary by country/company.
Software	A computer program.
Square Root	A number which, when multiplied by itself, generates the number of which it is the square root, e.g. 2 x 2 =4; 2 is the square root of 4.
Squirrel Stores	The unofficial hoarding of parts outside of the official stores system. Typically these parts are uncontrolled.

Stock	Items that are held in the expectation of future sale or use. Also referred to as *Inventory* and *Stores*.
Stockholding Costs	The combination of costs that are incurred as a result of owning inventory. Typically these include interest on the funds invested; labor to manage, clean, count, maintain, move, and otherwise process the inventory; buildings in which to house the inventory; and obsolescence or spoilage of the inventory.
Stock Keeping Unit	See SKU
Stock Out	Occurs when there is demand for an item, but no inventory of that item in stock.
Stock Over Maximum	Any inventory where the quantity held is in excess of the set maximum levels. Typically used in the SAP software.
Stock Turn	Stock turn measures inventory efficiency by measuring the rate at which the inventory is used. The stock turn is calculated by dividing the dollar value of stock issued in a year by the dollar value of stock held. For example, an item with annual demand of $1000 and a stock holding equal to $500 has a stock turn of 2.
Store	A place that is used to hold inventory. Also referred to as a *Warehouse* or *Storeroom*.
Stores	Items that are held in the expectation of future sale or use. Also referred to as *Inventory* and *Stock*.
Structure	Used to define organizational relationships within a company.
Supply Chain	The series of companies and activities that must

interact in order for an item to be created, moved, stored, sold/used, and ultimately delivered to the end user.

T

Tolerance for Risk — The ability of an organization to accept the risks associated with not holding stock.

Total Cost Ratio — The combined total percentage of finance and other holding costs for inventory.

Trading Stock — All raw materials, WIP, and finished goods that are intended for production and sale. Sometimes referred to as *Operating Stock*.

Trigger Point — The level of inventory at which replenishment is triggered. Also referred to as the *reorder point*. See also *Minimum*.

Turnover — This has two general uses. In inventory, it is similar to stock turn (see above). It is also used to describe the sales revenue of a company, e.g., Company XYZ has a turnover of $20M.

U

Unique Spares — A spare which is only used on a specific asset and which is depreciated over the life of that asset. Sometimes called a *Depreciable Spare*.

Unit Price — The price of a single item, even though the items may be bought in multiples.

Utilities — Gas, water, and electricity charges.

V

Vendor Managed Inventory — Inventory where the vendor decides on appropriate quantities and replenishment, but our company owns the inventory.

W

Weeks of Stock
: The number of weeks supply held of an item, based on the average demand per week, e.g., when an item has 10 demands per week on average and we hold 100 items, then we have 10 weeks of stock

WIP
: See *Work In Progress*.

Work in Progress
: Inventory that is partway through a manufacturing process. It is no longer raw material, but not yet finished goods.

Working Capital
: The funds invested in a company's cash, inventory, and accounts receivable. Net working capital equals current assets minus current liabilities. These are the funds that are required to finance the purchase of raw materials, conversion into finished goods, and the time until paid by customers.

Written Down Value
: The book value of an item after depreciation or other non-cash costs are taken into account. For example, an item that cost $1000 but has $400 of depreciation has a written down value of $600.

Z

Zero Inventory Mindset
: An approach to inventory management that seeks to minimize the investment in inventory and requires that all inventory above zero be justified.

Appendix B
Data Collection Questions

Organization

1. How many stores locations are there in this company / on this site?
   - Check for different businesses/divisions and different store types.

2. How is inventory classified and organized?
   - Machine/process
   - Type: mechanical or electrical
   - Active, inactive, obsolete
   - A, B, C

3. What are the organizational structure and responsibilities for inventory?
   - Draw out the organizational chart—to whom does this function report?
   - What is the relationship with purchasing/supply?
   - What are the responsibilities for both financial and customer promise outcomes?

4. What training in stock management have the inventory personnel received?

5. Is the staff remuneration linked to inventory performance? How?

Management and Control

6. How are items added to inventory and how is the recommended holding determined?
   - Max–Min
   - Reorder point

- Reorder quantity
- Safety stock
- History

7. What process is used to review stock holding targets and when were the targets last reviewed?
    - Cycle count
    - Order by order
    - Usage analysis

8. Is there a quality system for inventory management?

9. Does your computer system only record stock movements or is it also used to control stock quantities?

10. What approach is used to value stock
    - Average
    - Latest price
    - FIFO/LIFO

11. How are the stock level and availability controlled?
    - Max–Min
    - Visual management
    - Cycle count
    - A,B,C classification

12. How is obsolescence managed?

13. How are item movements recorded?
    - Is this system followed?

14. Who has access to the store?

15. What KPIs are used to monitor stores performance?
    - Availability/service level
    - $ Value
    - Turnover
    - Stock turns

16. What is the current value of these KPIs, actual vs. budget?

17. How and to whom are these KPIs reported?

18. How is the availability policy set?

19. What current other plans are there to reduce inventory?
    - Supply chain initiatives
    - Stores rationalization
    - Spares rationalization

Activity Data

20. What is the annual transactional value by store and category?

21. What is the current value of stock by store and category and how does this compare to budget?

22. How many purchase orders are raised per day or week for inventory items?

23. Is there any planned inventory held as stock build for project or marketing initiatives?

24. What benchmarking data do you use?

Reordering Process

25. Are orders 'rolled up' in order to manage purchasing?

26. Are local suppliers identified and/or favored?

27. Are orders shared with other stores?

28. What is the process for reordering and restocking stores items?
    - Need to understand the internal lead time
    - Need process and times from 'flag to restock'

Index